DATE

THE PLANIVERSE

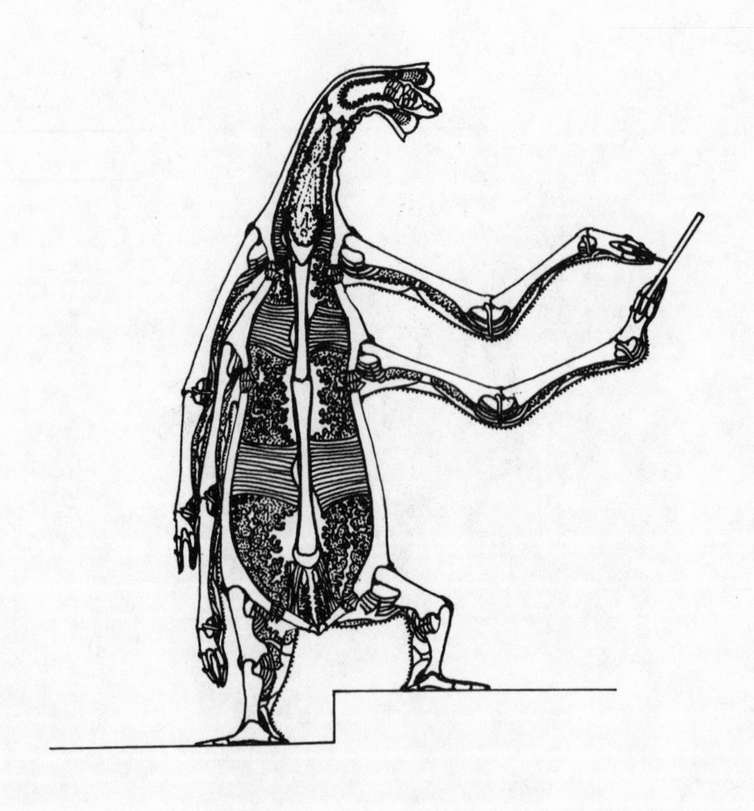

Computer Contact with a Two-Dimensional World

A. K. Dewdney

POSEIDON PRESS ▪ NEW YORK

A Poseidon Press Book
Published by Pocket Books,
A Division of Simon & Schuster, Inc.
Simon & Schuster Building
Rockefeller Center
1230 Avenue of the Americas
New York, New York 10020

Designed by Eve Kirch
Manufactured in the United States of America

10 9 8 7 6 5 4 3 2 1

Library of Congress Cataloging in Publication Data
Dewdney, A. K.
The planiverse.
1. Hyperspace. I. Title. II. Title: Two-dimensional world.
QA699.D48 1984 530.8 83-24799
ISBN 0-671-46362-4
ISBN 0-671-46363-2 Pbk.

Dedicated to my father
Selwyn Hanington Dewdney
artist, writer, scholar
1909–1979

2DWORLD

The following book, of which I am not so much the author as compiler, originates with the being whose picture appears in the title page. His name is Yendred and he inhabits a two-dimensional space I call "the Planiverse." The discovery of the Planiverse, a reality still not accepted by many people, makes an interesting story in itself, and the purpose of this introduction is to tell it.

The first contact with Yendred was made on one of our computer terminals only a year ago. We, my students and I, had been running a program called 2DWORLD, the result of several consecutive class projects carried from one term to the next. Originally designed to give students experience in scientific simulation and in large-scale programming projects, 2DWORLD soon took on a life of its own.

It started as an attempt to model physics in two dimensions. For example, a simple two-dimensional object might well be disk-shaped and be composed of billions of two-dimensional atoms.

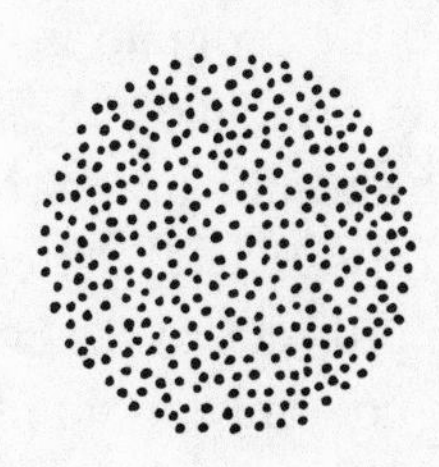

It has a kind of mass (dependent on the type and number of atoms in it) and may move about in the two-dimensional space represented by this page. As such, this space has no thickness and the disk is doomed to eternal confinement therein. We may further imagine that all objects in this space obey laws like those which prevail in our own space. Thus, if we give the disk a gentle push to the right, it will move to the right and continue moving to the right, sliding along at the same speed in a space which extends this page beyond your present location. Ultimately, within this imaginary plane, it will move far from the earth itself—unless, of course, it encounters a similar object:

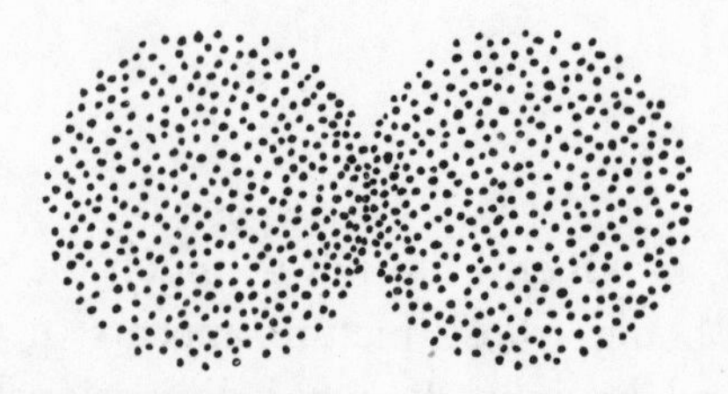

Doing so, the objects meet and suffer what physicists call an "elastic collision." Here we see the two objects at the precise instant of their greatest compression just before springing apart in opposite directions. In accordance with another well-known law in our own universe, the two disks carry off, between them, exactly as much combined energy as they brought to the collision. They have to collide, of course, since they cannot slip sideways past one another; there is no room in two dimensions for that, since no "sideways" is available.

This sort of thing is not difficult to represent on a computer graphics display terminal, nor is it difficult to write a program which creates all the behavior I have just described in the case of the two disks. Naturally, if there are to be atoms in the disks, there is considerably more work for the programmer to do and, when the program is running, considerably more work for the computer. But, in principle at least, one can write this sort of program and actually view the results on a screen.

The program called 2DWORLD started in a manner not unlike this. In the first term that the project was assigned, the students, under my direction, had incorporated not only a simple set of objects and laws governing momentum into 2DWORLD but had extended it to a system of planets orbiting a single star. At the end of the first term, there was talk among the students of a particular planet which they called "Astria." They wished to endow it with a geography of sorts and have beings called "Astrians" living on it. I nipped that ambition in the bud by reminding them that the term was nearly over and that examinations would soon be upon them. Besides, that sort of detail was mere wishful thinking and far beyond their abilities to embody in a program. Nevertheless, I counted the project a great success and congratulated myself on being clever enough to confine the dimensions of the physical space to two. It had, perhaps, made all the difference to the feasibility of the simulation program.

What had been especially stimulating for the students in this project, I believe, was the interaction of their other areas of knowledge with computing. They had been expected to read and digest a fair amount of physics and engineering. Moreover, just to give them a feeling for how a whole might be orchestrated from these parts, I had them read Abbott's *Flatland,* Hinton's *An Episode of Flatland,* and Burger's *Sphereland.* These works of fiction, or "science-fantasy" one might call them, each set up a model two-dimensional universe inhabited by two-dimensional beings. In this additional reading of the students lay the seeds of future development.

During the following term, I had wanted the class to construct programs which modeled predator-prey relations, among other things, but interest in 2DWORLD continued to run high and I finally gave in to their demands to continue the project. In that term we got to work on Astria, defining its surface of land and water and even attempting to construct simple organisms living in the water. Some of them could "eat" the others, so here at last was my opportunity to simulate predator-prey relationships, albeit not in the manner I had originally intended. I must admit that at this point I too was becoming rather more interested in the project than a certain sense of professional detachment would dictate. Frequently, we would stay well beyond the weekly laboratory period watching those little creatures glide and dart about in their two-dimensional realm, feeding and being fed upon.

But then came the end of the term and examinations for the students. I thought I had heard the last of 2DWORLD, at least until the autumn, but shortly after examinations were over, three of the keener students in my simulation class came to visit me. They wanted to continue working with 2DWORLD and were willing to stay in town over the summer to make that possible. One of them, a fourth-year student in biology, Winston Chan, wanted to design more advanced creatures for the simulation. The other two were computer science students, one of whom, Hugh Lambert, wanted to construct a "flexible, multipurpose query system" which would enable one to communicate with the 2DWORLD program as it ran. The other computer science student, Alice Little, had an idea which would allow 2DWORLD to be implemented on a much more elaborate scale while at the same time enabling it to run even faster. Faced with their obvious enthusiasm, I felt I had no choice but to give in to their requests. The computer they wanted to use would be largely free over the summer and, yes, I would be happy to continue coordinating the project.

Looking back over those summer months, I now realize that if 2DWORLD had a turning point, it was then. Time and again I came upon the three students working in the computer laboratory. This was a room housing one of the newer large-scale computers linked to a graphics terminal in the room and communicating with numerous other terminals throughout the building. One of the students would be at the terminal experimenting with some new program concept while the other two would be bent over a large worktable in the middle of the room, muttering to each other above the reams of printout spread thereon. Occasionally I would enter the laboratory and ask if they were having any problems. Politely, they would show me what they were doing, but I almost invariably ended up feeling more like a spectator than a participant. Nevertheless, I was fascinated with some of Chan's creatures and impressed with Little's new simulation system: her idea was to "focus" the activity of the program in a small area of the two-dimensional universe being simulated, allowing approximate behavior outside the area. Lambert worked closely with Little in developing a technique for verbally querying the system to discover its "background behavior"—whatever was not obvious from looking at the display screen.

When the summer ended, I fully expected the three students to go their separate ways. For one thing, they had all graduated and two were ready to enter high-paying jobs in the computer industry. But Little had decided to do graduate work in computer science and Lambert wanted to work in the department as a course assistant. This development alarmed me a bit, especially when both applied as assistants in the Advanced Simulation course I was giving. Choosing Little, because of her more obvious commitment to an academic career, I was surprised to discover that Lambert would be auditing the course. I was more than a bit alarmed to think that the course would be dominated by 2DWORLD once again.

I needn't have worried: all the incoming students had heard about the project and virtually refused to work on anything else. Obviously, 2DWORLD was meeting some deep, emotional need of the students. Many of them seemed determined to retain some fragment of their childhood fantasy life. One had only to think of the immense popularity of various computer games like "Space War" and "Adventure" to understand what a widespread phenomenon it was. Nevertheless, I was determined that the course should have some semblance of academically respectable content, so I spent the first two weeks on a perhaps overly learned series of lectures about generating random numbers.

We eased into 2DWORLD rather slowly. By now it had become a very large and rather sophisticated program of several thousand instructions. It took Alice Little and me several weeks to initiate the new students and to find subprojects for them which somehow fitted into 2DWORLD as a whole. Chan had left us with a rather strange ecosystem: the oceans now held several species of plant and animal life, the former being nearly as carnivorous as the latter. Apparently, two-dimensionality made the carnivorous plant business much more profitable than in our world. Chan had, moreover, invented an animal called a "throg" which slightly resembled a frog. It apparently did nearly as well on land as in the ocean. On other fronts, Alice's new simulation management system worked wonders for our ability to include new detail with only slight increases in computation time. Perhaps this factor as much as any other was what prevented 2DWORLD from becoming totally unwieldy during that crucial second year.

Just before the Christmas break, I invited Chan in to lecture the class. As a hobby, he had been laying down the design parameters for a new class of land animals, including one which was supposed to be "intelligent." I don't believe that at the time Chan had any very clear idea of how such intelligence would be built in, but two of the students took up Chan's ideas, changing projects in midcourse, so to speak. These two, Ffennell and Edwards, produced an upright, vaguely triangular creature (called an FEC) which moved about on its base by alternately extending and retracting two pods and rocking as it went. Its intelligence, as such, did not reside in a simulated brain. There was, after all, a limit to how much detail even Alice's system could handle. Instead, its intelligence was embodied in a special subprogram modeled after certain recent work in artificial intelligence. To round out the picture before our narrative enters its next, rather dramatic phase, it seems only fair to mention that other students concentrated on Astrian weather, atomic physics, biochemistry, ecological balance, and so on. These projects were perhaps not quite as sophisticated as they sound, but each contributed something interesting and vital to 2DWORLD for the most part.

As might have been expected at this point, 2DWORLD had become something of a sensation at our university. Not only had the campus newspaper done two stories about us, but the dean of science had made a special point of visiting our laboratory. I have saved the printout from that occasion. Lambert had been running a demonstration at the graphics terminal, typing queries to the system while one of the Ffennell-Edwards creatures trundled awkwardly across the screen in search of throgs.

```
▪ RUN PRIORITY QUERY.
  READY.
▪ VERBAL: FEC IN FOCUS.
  FEC HERE-HUNT THROGS.
▪ WHY?
  UNKNOWN: ''WHY.''
```

At this point, I recall, Lambert explained to the dean that the word "why" was not in the system vocabulary.

```
▪ DESCRIBE HUNT.
```

```
MOVE EAST UNTIL THROG OR SEA.

▪ DESCRIBE UNTIL.

UNTIL INCLUDES AVOID AND RETURN.

▪ DESCRIBE AVOID.

AVOID INCLUDES WATCH FOR AUNT LYONS.
```

Although the dean was highly impressed on one level, I could tell from the faintly anxious look which occasionally crossed his features that he, as I, had some reservations about the academic usefulness of the 2DWORLD project. A moment after the printed conversation, a throg hopped into the focus, became rather detailed, then sensed the FEC and tried to hop away again. The FEC quickly leaned over, extended its eastern arm, and trapped the creature between body and ground. Scraping its arm along the ground, the FEC managed to catch the throg between its two fingers. The dean gaped as the FEC lowered the throg into its mouth and chewed it, gradually, into little pieces. Being two-dimensional, the inside of the FEC was just as visible as its outside: the pieces of throg made their way, one by one, into a digestive pouch just below the jaws.

During the next term, 2DWORLD developed to the point where even Alice Little's management system could handle no more additional features. There was now an atmosphere on Astria, regular patterns of weather fluctuated above its surface, and quite a variety of creatures inhabited both the ocean and the land. The FECs, moreover, had been given simple underground dwellings where they could store food and have inane conversations about digging or throg-hunting.

One night, near the end of the term, I returned to my office after supper to prepare a talk I was giving at a Data Structures conference in a few days' time. In the middle of a period of deep concentration, there came a light, fluttery knock at the door and in stepped Alice Little, her face looking strained and puzzled.

"Dr. Dewdney, something's gone wong with the system! One of the FECs said a word not in its vocabulawy. We checked the dictionawy but no one has changed it!"

She seemed breathless, and her speech impediment, barely noticeable under normal circumstances, was now quite pronounced. I was annoyed to have my train of thought broken but put on my good-natured teacher's face and rose from my chair. I tried to reassure her as we walked down the corridor to the laboratory.

"The word is Y-N-D-R-D." She spelled it out, being careful with the R.

"That doesn't sound like one of our words."

"I know!"

All the lights were off in the laboratory. The graphics scope bathed the room in an eerie radiance as we sat down at the terminal in front of it. A student named Craine who had been working with Alice shifted his chair to make room and we three stared at the lone FEC on the screen. It swung its head slowly to the right and then to the left.

▪ RUN PRIORITY QUERY.

READY.

▪ VERBAL: FEC IN FOCUS.

FEC HERE-YNDRD.

"There it is," said Alice under her breath. We spent a few minutes discussing how we might track down the problem, then I took over at the terminal.

▪ DESCRIBE YNDRD.

YNDRD IS BUT FEW WORDS.

Craine murmured "Oh, no!" and continued to do so every few minutes thereafter. This did not help our analysis very much.

We reasoned, after repeating this sequence a number of times to rule out spurious errors, that everything had to make sense in the context of the simulation. The phrase "is but few" reflected a construction unavailable to the program and we decided, finally, that it represented two separate sentences, "Yndrd is" and "But few words."

▪ WHERE IS YNDRD?

There was a long pause.

NO WORD.
▪ DESCRIBE WORD.
WORD FOR WHERE YNDRD IS.

This was getting a bit strange. The FEC would normally have described its position relative to nearby landmarks on the planetary surface. Perhaps there were none.

"Alice, where's the nearest feature that an FEC would relate to?"

Learning that there was another FEC just to the west of this one, I asked Alice to have it come east into the focus area. As it came on the screen, Alice typed:

▪ EAVESDROP.
WHO ARE YOU?

I'M GEORGE. WHO ARE YOU?

I'M CHAN THE MAN.

Thus ran the first three sentences of conversation between the FECs. I took it that "George" and "Chan the man" were two of the regular FEC names, invented by the students last term. It seemed the glitch had disappeared.

"I *thought* that was George all along!" declared Alice with an air of relief. Craine agreed solemnly. I decided to interrupt the conversation going on before us.

▪ VERBAL: EAST FEC IN FOCUS.

FEC HERE—CHAN THE MAN.

▪ WEST FEC IN FOCUS.

FEC HERE—GEORGE.

▪ DESCRIBE YNDRD.

UNKNOWN: ''YNDRD.''

I spooled off the printout, our only documentation of the incident, tore it from the terminal, and took it back to my office, suggesting that Alice and the student call it a night. In my office, I studied the printout for a few minutes before succumbing once again to the pressure of my imminent conference talk. There was really no way that sort of thing should be happening.

A few days later, I spied Edwards and Craine in the cafeteria and went to their table. I found them discussing the "yendred incident," news of which had already leaked out to the rest of the class. Edwards, evidently, was taking the "but few words" statement at something more than face value.

"How do *you* explain it?" they asked me.

I chose my words carefully. "Well, I don't know enough about the 2DWORLD system as a whole to say just what combination of software and hardware errors could produce a sequence like that. Perhaps no one person does. It's certainly become very complex. On the other hand, you might start by analyzing the ASCII code for YNDRD and look for a systematic bit error that would produce it from any of the five-letter words currently in our vocabulary."

Craine showed real interest in this suggestion, but Edwards gazed abstractedly out the cafeteria windows.

On May 22 a telephone call interrupted our family supper. YNDRD had reappeared at our terminal.

"Professor Dewdney?"

"Yes. Is that you, Alice?"

The details of the recurrence, as related in a few breathless sentences by Alice, were sufficiently startling to put an end to my supper. I promised to drive to the university right away.

In the laboratory there was quite a crowd: Alice, Chan, Lambert, and several students. The lights were on and everyone seemed to be talking at once. On the display screen a lone FEC rocked gently from side to side. Flowing from the terminal and arranged into a carelessly folded pile was a large mass of printout. My entrance created a sudden silence.

"What, exactly, is going on?"

Edwards strode to the terminal and picked up part of the printout. "See for yourself, sir." He proffered a fragment.

▪ DESCRIBE YNDRD.

YNDRD YOU THERE ARE KNOWS BUT HE YOU DOES NOT SEE AND YOU DOES NOT HEAR.

▪ MY NAME IS CHAN.

UNKNOWN: ''CHAN.'' DESCRIBE YOU.

▪ I AM A STUDENT.

UNKNOWN: ''STUDENT.'' DESCRIBE YOU.

The conversation went on for several pages, becoming steadily more unbelievable. I soon came upon a section in which a number of new words appeared.

▪ WHERE IS YOUR PLACE?

MY PLACE ARDE IS. WE THE NSANA ARE.

▪ ARE YOU TWO-DIMENSIONAL?

UNKNOWN WHAT IS ''TWO-DIMENSIONAL.''

Edwards interrupted to point out that the system queries and replies no longer followed the proper format. It would seem that the system now responded to a great variety of English sentences. It went through my mind that at any moment all the students would break out laughing at their wonderful joke on me. Further on I found:

BEFORE YOU MORE WITH ME WERE. NOW YOU LESS ARE.

▪ RUN SYSTEM.

READY.

▪ FOCUS (SAME, ½). VERBAL: FEC IN FOCUS.

YOU EVEN LESS MORE WITH ME ARE.

Alice, who had been looking over my shoulder, explained. "Here's where we started to lose him. It took us a long time to find out that we had to slow down the rocking motion."

"What has that got to do with it?"

She didn't know.

The rest of the printout, some twenty more pages, was no less strange. There did appear to be an entity conversing with the students, exchanging words with them and becoming continually more fluent in the process. The entity, YNDRD, alternated between describing a place called ARDE and questioning us about our world. I tried to keep my mind clear. Only two alternatives appeared possible: either the program was producing this wonderful conversation all by itself, or some remote prankster had a line to our terminal, perhaps someone in Dr. Barnett's Operating Systems course.

"Has anyone checked the hardware for some kind of remote patch?"

Alice replied, in an almost motherly way, that all inputs to the computer, except for this one terminal, had been disabled. They too had thought of this possibility. Numbly, I sat down at the terminal.

▪ DESCRIBE YNDRD.

YOU CHAN ARE?

▪ NO. I AM DEWDNEY.

WHY NO ONE FOR A LONG TIME TALKED?

▪ WE WERE TALKING TO EACH OTHER.

WHAT ABOUT TO EACH OTHER WERE TALKING?

A dry, metallic taste invaded my mouth and I felt somewhat lightheaded.

▪ WE TALKED ABOUT YOU.

YOU FROM THE BEYOND ARE?

▪ YES. IN A MANNER OF SPEAKING.

YOU SPIRITS ARE?

▪ WE ARE NOT WHAT WE OURSELVES WOULD CALL SPIRITS. WE HAVE SOLID THREE-DIMENSIONAL BODIES.

WHAT ''THREE-DIMENSIONAL'' IS?

I stayed on with the students that day. We spent several more hours talking with the being we had come to acknowledge as "Yendred," generating more reams of output in the process. As the conversation continued, strange things began to happen on the screen: Yendred's anatomy was slowly changing from that of an FEC to something completely beyond our imaginations. Strange internal organs pulsed rhythmically. Sheets of tissue parted along seams and then rejoined. Our FECs, moreover, had two arms. This creature had four, and they seemed to inflate and deflate regularly. Adjacent to Yendred, curious plants appeared, hugging the Ardean surface. The entire scene was now far more detailed than anything produced by the 2DWORLD simulation program. It had, moreover, an odd, alien quality, utterly convincing because it was so bizarre.

In spite of this transformation in the two-dimensional landscape before us, both the scanning program and Alice's focusing system continued to work, so that we could scan the landscape at will and examine any part of it in some detail.

When we at last said farewell to Yendred and terminated the 2DWORLD program, we sat in silence for nearly a minute, each of us absorbing the impact of this astounding manifestation. Present were Chan, Edwards, Ffennell, Lambert, Little, Craine, and, of course, me. I made a speech which went something like this:

"What we have been witnessing just might be real. That is, it just might represent a communication with beings in another universe. If it's a prank, someone has gone to an *enormous* amount of trouble to entertain us. In any case, I think we had all better agree right now to keep this evening's events a secret until we understand what's going on a little better."

Lambert interrupted. "I think we should have a public demonstration."

"Oh, be serious!" said Chan.

"I'm afraid a public demonstration is out of the question: either way we lose. If people thought it was real, we would quickly lose control of the thing. Before you knew it, this building would be full of astronomers, anthropologists, and God knows what from every corner of the globe. You would certainly get a lot of publicity as creators of the 2DWORLD program, but you'd never get to use it again. On the other hand, if people thought it was a prank or a hoax, my reputation—and yours—would suffer very badly. I could even be pressured to resign my position here, and we would almost certainly lose our access to the facility.

"But suppose we didn't mind letting a task force of scientists take over. What do you think the result would be? Imagine each specialist lining up at the terminal to ask his or her little question of Yendred. Imagine the bureaucracy involved. If you were a two-dimensional creature, if you were Yendred, would you rather deal with a small group of sympathetic and intelligent people or an endless stream of cold-blooded inquisitors? I mean, can you imagine

```
HELLO YENDRED. IM DOCTOR PIFFLEWHIZ FROM
HUMBUG UNIVERSITY. MY SPECIALTY IS PSYCHO-
HYDRODYNAMICS AND I WANT TO KNOW IF YOUR
PEOPLE EVER GET WATER ON THE BRAIN.
```

"No. We really have no choice in the matter. We *must* keep this a secret, every one of us. I'm not even going to tell my wife!" I looked at the individual faces before me. "Agreed?"

They nodded solemnly.

Succeeding contacts with Yendred, some of which lasted from eight to ten hours, always seemed to begin when one of our FECs was in a solitary state on the Astrian surface. We would continually query the FEC, waiting impatiently as it explained how to catch throgs. Suddenly Yendred's words would appear in place of the FEC's responses and the scene would slowly transform itself into Yendred's current environment on Arde. With succeeding contacts, this phenomenon took place with increasing speed. It seemed to have much to do with Yendred's own mental state: he once told us that he merely had to contemplate the "space beyond space" in order to feel our "presence" and bring his own world once more into coincidence with the 2DWORLD simulation program.

During that summer, we built up an extensive library of conversations with Yendred. I had an extra bookshelf brought into my office and we began to accumulate printout paper at the rate of two or three large file folders every contact.

The students and I met once a week, discussing the past week and deciding what topics we would question Yendred about during the next session. For his part, Yendred would often take us to some nearby feature, directing our attention to various details and explaining them to us. I don't think he ever fully grasped the comprehensiveness of our vision. He would assume that something normally invisible to him (say, the inside of an Ardean steam engine) was also invisible to us. But it wasn't. At other times, rather than explain an adjacent object, Yendred would discuss astronomy or animal behavior or Ardean morals or the Ardean monetary system, things for which pictures would not be of much use in any event. For the rest, he would sometimes question us at great length about human beings, the earth, and our universe.

Before the second week of contact with Yendred had quite elapsed, he left his home and began a journey which, though interrupted several times, continued for two months. The object of this journey is rather difficult to define, having something to do with what Yendred called "knowledge of the beyond." In any case, we were very fortunate that he had chosen this particular time to travel, as we saw a good deal of Arde, its people, technology, and culture as a result. It is also just possible that Yendred's preoccupation with the "beyond" provided a unique mental set, making him the favored target of our simulation program—the nexus of Arde.

Through all of this, none of us knew how long these contacts would last. There was a certain anxiety in the air: what had appeared so suddenly and mysteriously could vanish in the same manner.

In the meantime, things were not going very smoothly in the world outside our laboratory. Rumors circulated about strange goings-on in the 2DWORLD project. Indeed, a second wave of publicity quickly developed. It centered on a student who claimed that we had "contacted another world." All the university needed was the following headline in a well-known tabloid with international distribution: PROFESSOR DISCOVERS FLAT WORLD.

Within a few days a message came directly from the university president to our chairman. I was to put an end to all my experiments at once. The chairman, in his kindly way, said, "Crikey, man, whatever you're doing, stop it!"

Torn between loyalty to the university and curiosity about Arde, I finally resolved to continue our "experiments" during secret sessions in the small hours of the morning. To avoid attracting attention, we agreed to enter and leave the laboratory singly and to keep the lights off during the contacts.

The sessions continued until the morning of August 4, when they abruptly ended. In previous sessions, Yendred had been traversing a portion of the high central plateau on Arde's continent and had encountered a fellow Ardean, Drabk by name, not far from a landmark connected with an ancient Ardean religion. After a number of meetings with this other being, Yendred announced that his journey would continue no farther for the present. Up to this point, it would not be presumptuous to claim a certain sense of "friendship" with Yendred: he had always willingly shared information with us and had shown what can only be described as enthusiasm in learning about Earth and its inhabitants. But in the next to last session, he had displayed few of these qualities and had terminated the contact almost as soon as it began, claiming some need for secrecy in the proceedings.

In the last session, Yendred spoke with us for several hours in Drabk's absence. Just before Drabk appeared, however, Yendred underwent a remarkable change. Yendred then spoke no more with us until the very end of the contact.

WE CANNOT TALK AGAIN. TO TALK AGAIN IS OF NO BENEFIT.

▪ BUT WE HAVE SO MUCH MORE TO LEARN FROM YOU.

YOU CANNOT LEARN FROM ME. NOR I FROM YOU. YOU DO NOT HAVE THE KNOWLEDGE.

▪ WHAT KNOWLEDGE?

THE KNOWLEDGE BEYOND THOUGHT OF THE REALITY BEYOND REALITY.

▪ WOULD IT HELP IF WE LEARNED YOUR PHILOSOPHY AND RELIGION?

IT HAS NOT TO DO WITH WHAT YOU CALL
PHILOSOPHY OR RELIGION. IF YOU FOLLOW
ONLY THOUGHT YOU WILL NEVER DISCOVER
THE SURPRISE WHICH LIES BEYOND THOUGHT.

▪ WHAT SURPRISE?

UNKNOWN: ''WHAT SURPRISE.''

▪ FEC IN FOCUS.

ADOLF HERE—WAITING FOR THROGS.

From that point on, attempts to raise Yendred resulted only in the by now usual UNKNOWN: "YNDRD" message. Although there has been a certain amount of publicity resulting from our contact with Arde and the Planiverse, no one seriously expects the link to be reestablished. As a result, interest in the 2DWORLD program has fallen back to its old level. Alice, currently finishing her graduate work, sometimes loads the program from tape to disk and plays with it for a while. I have no doubt that she hopes to contact Yendred again, and I have no doubt that she never will. She will never come breathless to my office door again, presumably, because we are a people with no "knowledge beyond thought," whatever that means. Nevertheless, I have encouraged Alice to write a description of the 2DWORLD program (omitting all references to Yendred and the Planiverse) for publication. Perhaps one of the popular computer magazines would be interested.

For my part, I have already taken some time away from university life to document the two-dimensional universe revealed by our "experiments." Essentially, I have edited all the transcripts into a set of more or less coherent chapters, each devoted to a separate contact period. I should qualify the word "edited" here because only small segments of the transcripts are ever quoted directly; the rest have been condensed into summaries and descriptive accounts. Some of the more technical details, especially matters of science and technology, have been set aside in special boxes which nontechnical readers may wish to skip—at least on their first time through the book. I prefer to think, however, that many of these will be intelligible to such readers.

It would not be possible to close this already lengthy introduction without some attempt at an explanation of what was happening last summer. First, I must say what is probably already obvious to the reader: I take Arde seriously and believe that it exists. Where? Not in our universe, I think, and therefore not anywhere in particular in relation to us. My only guide in this matter has been a certain familiarity with the vast world of conceptual models comprised by modern mathematics. Not only does there fail to be any convincing reason why our universe should be the only one, but there seems to be no reason why other possible universes should not have two physical dimensions instead of three.

How, then, did Yendred appear in our computer? This question has taxed me to the limits of my imagination and the only conceivable answer is based on what might be called a "theory of coincidence," an analogy, if you like, to what happens when a vibrating tuning fork is placed beside a nonvibrating one: if the tuning forks are the same size, the quiet one will begin to vibrate also. The two forks do not have to be made from the same material or even to have exactly the same shape. If one could in some way write down all the information which is contained in those two forks, one would find the information coinciding in a certain small number of important respects and somewhat diverging in all others. In particular, the information describing the vibrations of the first tuning fork is, in a sense, absent from the description of the second tuning fork. Forgetting for a moment about the theory of sympathetic vibrations, imagine instead a mysterious tendency for the information missing from the second description to be replaced by the vibratory information of the first description.

Perhaps something like this happened when the 2DWORLD simulation program reached a certain level of complexity. Our "Planiverse" coincided with Yendred's two-dimensional universe in enough important respects that the latter set the former vibrating, so to speak:

YNDRD IS BUT FEW WORDS

Even Yendred's native language was translated into English, albeit with curiously scrambled sentences. Did the Ffennell-Edwards artificial intelligence program correspond in some way to Yendred's mind? Did Lambert's query system somehow match the Ardean speech center?

One may speculate endlessly about the actual mechanisms involved, but being very much out on a limb, philosophically speaking, I will stop here. It is useless to speculate further.

Wytham Abbey
Wytham, Oxford
June, 1981

1

Arde

Yendred lived (and still lives, I am sure) upon a disk-shaped planet called Arde. At the beginning of our contacts we knew nothing about Arde, but by the end we knew a great deal. This short chapter presents a certain minimum of information necessary to set the stage for the chapters that follow; we would have been very grateful for such a description in the first few weeks of our contact with Yendred!

The planet Arde has a hot, molten interior and a cool, hard crust supporting a circular geography which is three-quarters ocean and one-quarter land. The ocean is called Fiddib Har, and the land is called Ajem Kollosh.

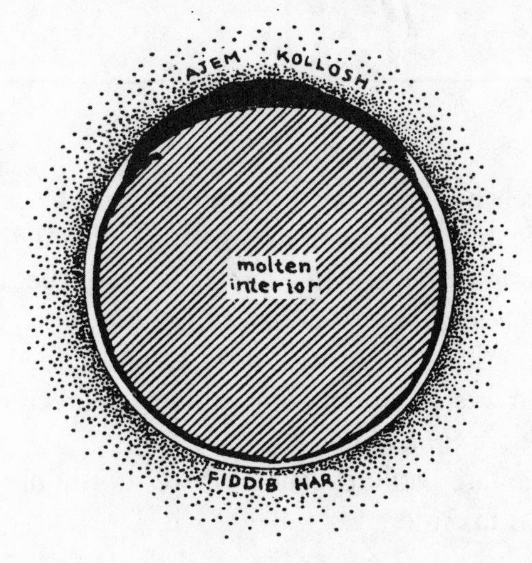

There is gravity in the Planiverse, and it is gravity which accounts for the planet's circular shape. Arde is really an enormous disk of two-dimensional matter pulled into this shape by the mutual gravity of all its constituent particles. Arde is circular for the same reason that Earth is spherical.

Besides its single ocean and lone continent, Arde possesses an atmosphere. Densest at the surface of the planet, it thins out with increasing altitude, fading away into the near-vacuum of two-dimensional interstellar space, the very fabric of the Planiverse. Far away in this space swims the star called Shems. Both Arde and a sister planet, Nagas, travel around Shems in orbits determined by the same unrelenting gravity which rules their circular shape.

The Ardeans (as we have come to call them) live on the continent of Ajem Kollosh. Rather, it is slightly more accurate to say that they live *within* it, for their homes are underground. Nevertheless, we may easily picture an Ardean standing on the surface of Ajem Kollosh.

The Ardean (which has been filled in entirely with black—a device for avoiding the depiction of its internal organs) is pointing east along the surface of the ground. We call this direction "east" because this is the direction in which Shems rises every morning. Ardeans call this direction "punl" and the opposite one (west) "vanl." On Arde there is no north or south, just east and west. The only other principal directions in which an Ardean may point are up and down.

Night and day are easily explained if we grossly distort the size of Ardeans in relation to Arde.

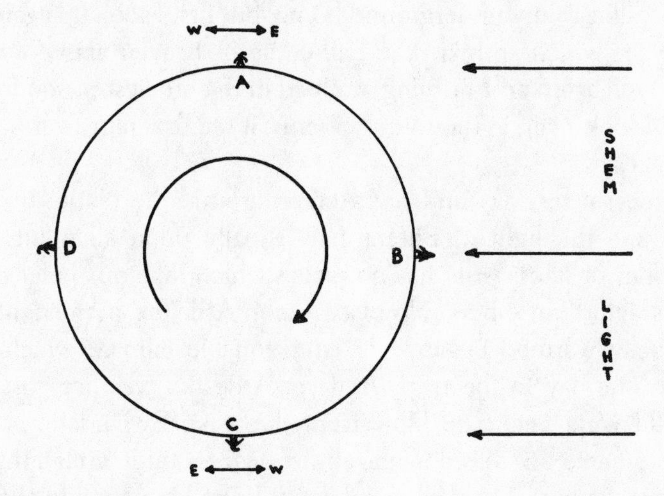

We have measured the Ardean day to be 31 hours and 25 minutes long. This is the time it takes Arde to complete one clockwise rotation. At the beginning of such a day, an Ardean at position *A* might well be looking east, admiring the splendor of Shems rising above the distant horizon. Approximately 7 hours and 51 minutes after this, Arde's rotation has brought the Ardean around to position *B,* where Shems is directly overhead. It is now noon. After yet another 7 hours and 51 minutes, daylight comes to an end for the Ardean at *C* as Shems appears to sink below the western horizon. For the next 15 hours and 42 minutes the Ardean will be plunged into night, passing the midnight point at *D.*

On Earth the length of daylight varies with the season, but on Arde night and day are exactly equal in length and never vary with the season.

I must confess to very strange feelings as I write about day and night on Arde. Especially when I write about sunrise and sunset I imagine myself to be an Ardean standing on some rocky summit and watching a sliver of sun set over a slip of land. All is confined to the narrowest band of vision imaginable; an infinitesimal line, encompassing me, contains my entire visual world. I turn my head from Shems setting in the west, turn my head upward (not around) and swing my crystalline eyes to the east where all is darkness and settling gloom. As an Ardean I watch this fading light, taste the evening, listen to the sounds of burrowing things underfoot, and am satisfied with the day which draws to a close; soon I will go down some stairs to my underground home, but first I look up again at the gathering stars which sprinkle the half-circle of heaven above my head. Suddenly, I am again an Earthling, trapped in the cruelest prison imaginable. When I look again at the setting Shems, it seems a narrow mockery of a sunset on Earth.

As Arde rotates, day and night creep around the planet in endless succession and the light of Shems falls equally upon all points of the surface sooner or later. Arde has no places which, like our polar regions, receive less light than others. No places upon Arde are permanently cold but all equally warmed by an even and equable climate which hardly varies from one day to the next. And yet Arde has weather: just before noon a mild wind begins to blow from the west. By midafternoon the wind is quite fierce but dies down again by degrees until, after sunset, it is once again a zephyr. Then all is calm until just before midnight, when another wind begins, this time from the east. It too builds in strength through the wee hours of the morning. Reaching a climax a few hours before dawn, it dies away just like the afternoon westerly. By midmorning, all is again calm for a time.

Arde has seasons, but these spring from an entirely different cause than our seasons on Earth. If Arde stayed within the same narrow limits of distance from Shems as Earth does from its sun, there would be no seasons to speak of on Arde. Seasons on Earth are due to the tilt of its axis toward the sun or away from it. Seasons on Arde are due to its rather wild orbit, a path which weaves a beautiful pattern around Shems, taking it twice as close during the summer season as it comes during the winter. In summer, the weather is warm and Shems appears large and bright. Occasionally rain falls, usually after sunset, but this does not amount to much. In winter, the sky grows progressively cloudier and the days are wrapped in a cool gloom. It rains frequently and the surface of Ajem Kollosh is scoured by rivers.

This brings us at last to Ajem Kollosh itself and a few pertinent features of Ardean geography. Arde's single continent is divided into three more or less distinct geographical regions. If we could remove all of Ajem Kollosh from Arde and lay it out along a straight line, these three regions would be immediately visible, at least if once again we could distort the scale of things.

DAHL RADAM

PUNIZLA

VANIZLA

The three regions are called Punizla, Dahl Radam, and Vanizla by the Ardeans. Punizla and Vanizla are low-lying, almost level regions to the west and east, respectively, of Dahl Radam, a high central plateau. The Ardeans are divided into two nations occupying Punizla and Vanizla, but Dahl Radam is almost uninhabited. However, the political boundaries of Punizla and Vanizla meet at a point near the center of Dahl Radam.

Until now I have put off saying how "large" any of these features are, or indeed, how large Arde itself is. Obviously, we have no way of directly comparing distances on Arde with distances on Earth, so we have adopted the simple convention that Yendred's height and my own (175 cm.) are the same. This has made it possible to obtain rough measurements of everything appearing on our display screen in "Yendred units," easily converted to metric.

By this measurement, both Punizla and Vanizla slope gently upward from their respective shorepoints approximately 1,000 kilometers before the grade begins to steepen significantly at Dahl Radam. The great plateau itself stretches roughly 800 kilometers.

Yendred is a Punizlan, and the coming chapters document a journey he made from the shore of Punizla to the heights of Dahl Radam. Starting in the early Ardean spring, he traveled this vast distance partly on foot and partly by other means of transportation. The winter rains had not yet ceased, and because these were a continuing factor in his earlier travels, it would be useful to indicate here and now some of the startling properties of these rains in connection with the two-dimensional surface of Arde.

Although it does not rain much during the late spring and summer on the low-lying parts of Ajem Kollosh, it rains every day on the slopes of Dahl Radam, winter and summer. The steep grade of these slopes illustrates even more dramatically what happens when it rains anywhere on Arde.

When the afternoon westerly blows up the western slope of Dahl Radam, moisture-bearing air becomes cooler and the water vapor begins to condense into clouds. The clouds increase in size and condensation until the water droplets become too heavy to be borne by the clouds, falling out as rain.

The slopes of Dahl Radam are covered by a very coarse "soil" consisting largely of rough pebbles and rocks. When the rain falls on this surface, it cannot penetrate below the ground as it does on Earth. For everywhere, the pebbles and rocks on the ground of Dahl Radam form a continuous chain of contact, a barrier which admits no water whatever.

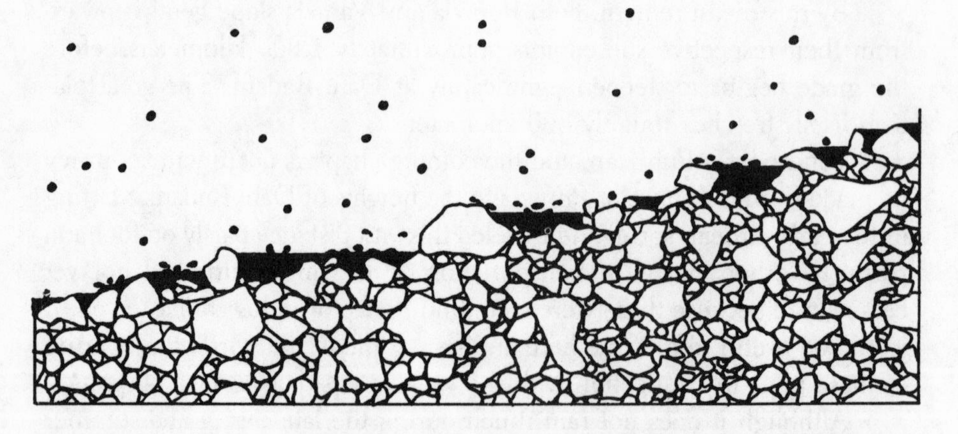

The same thing is true in the low-lying levels of Punizla and Vanizla, where the soil particles are much finer: it would then be as though we were seeing the above picture through a microscope.

The rain, unsuccessful at penetrating the soil of Dahl Radam, has no choice but to run downhill and, if the rain lasts a long time, the depth of the resulting "river" becomes significant and the force of its current is able to strip away a few layers of those rocks and pebbles which had earlier resisted its penetration. If by chance the river should encounter a large boulder, either this too will be rolled along by the river or a waterfall will form over it, excavating the ground on its downslope side until the boulder rolls into the hollow thus created.

By this and many other mechanisms, running water is the chief sculptural agent on the surface of Arde, wearing large rocks down to pebbles and pebbles down to fine soil. It moves rock from the top of Dahl Radam downward to the level reaches of Punizla and Vanizla, converting the rock to soil and producing, in consequence, the smooth contours of the great plateau. The same action in a gentler and more microscopic form has sculpted Punizla and Vanizla into their present linear topography.

Whether thundering down the slopes of Dahl Radam or flowing gently along the lowlands, the temporary rivers of Arde eventually come to an end. Gradually evaporating, they lose force and begin to deposit their burden of boulders or soil as they die. The only trace they leave is a long stretch of ground saturated with water to a certain depth.

And now the same continuous chains of contact which at first prevented the water from entering the ground prevent it from leaving. On the slopes of Dahl Radam, this is not a serious problem, but should one of the gentler rivers of Punizla or Vanizla end in a depression or pit, the result is a deadly quicksand into which the unwary Ardean may easily step. For the top layer of soil will dry out completely and present a firm-looking surface to the eye. Luckily, there are other ways in which the soils of Ajem Kollosh may eventually become dry—otherwise the Ardeans would have been drowned eons ago in a sea of mud.

The landscape of Ajem Kollosh is as dull and featureless a place as one could imagine. One would be hard pressed to understand how life is possible in such a place, and yet life is relatively abundant on Arde, as we shall soon see.

Many of the contacts about to be described contain details which were not learned at the time of the contacts but at some other time, usually later. This is especially true of some of the scientific and technical details appearing in these descriptions. Many of these were learned, for example, during Yendred's visit to a scientific establishment halfway through his travels.

A great deal of additional information on Ardean science and technology has been included in an appendix at the back of the book.

2

A House by the Sea

Yendred was hatched in a house by the Punizlan seashore some thirty Arde years before our first encounter with him. The house was not really "by the sea," since there were five other houses between his and the shore of Fiddib Har. Nor did it overlook the sea in any sense, since it was entirely underground.

Wednesday, May 28, 8:00 P.M.

When the simulation program had finally meshed with Yendred's world, we found him standing on the surface, his head pointing straight up in the posture which we would later learn was habitual with Ardeans in a day-dreaming or meditative state. To his east was a stairway descending to his house. We changed the scale on the display screen and obtained the following view of both Yendred and the house.

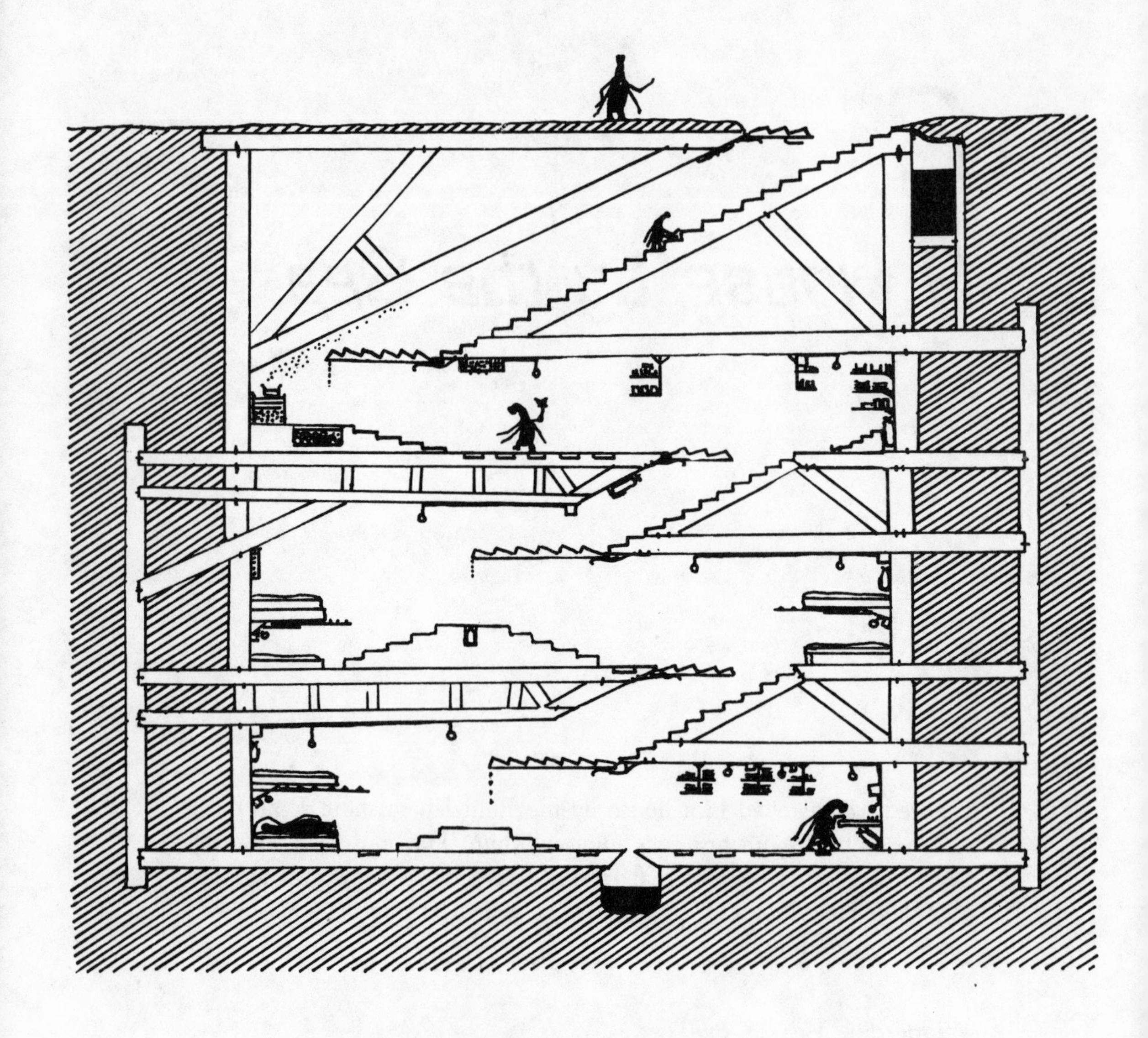

▪ WE SEE A HOUSE UNDERGROUND NEAR YOU.

THAT THE ENTRANCE TO MY HOUSE IS. I YOU INSIDE WILL TAKE.

▪ WE CAN SEE INSIDE ALREADY.

THEN YOU LIKE SPIRITS ARE. YOUR EYES SEE THROUGH ALL THINGS.

▪ NO. WE SEE YOU FROM OUTSIDE YOUR SPACE. A DIFFERENT DIMENSION.

I CANNOT THAT IMAGINE. IS IT A SPACE BEYOND SPACE?

▪ YES. BUT IN A DIRECTION YOU CANNOT POINT. WE CAN SEE INSIDE YOU AS EASILY AS WE CAN SEE INSIDE YOUR HOUSE. [Here, Yendred gave what appeared to be an involuntary shudder. He said nothing for several minutes.]

IF YOU INSIDE THE HOUSE CAN SEE, TELL ME WHO IS WITHIN.

▪ WE SEE FOUR NSANA. ONE OF THEM IS SMALL. WHO ARE THEY?

THOSE MY PARENTS GRANDMOTHER AND SISTER ARE.

We questioned Yendred about various details of the house and its occupants, learning in the process a great deal about everything from mechanisms to social life. In this as in previous contacts, he was slow to accept the extent of our "powers."

The house had three levels or floors, all arrived at by stairways, some of which were movable so as to avoid access problems. The top level contained a sort of kitchen and eating area, seats being collapsible and recessed into the floor when not in use. The second level had four beds, and the bottom level contained two beds at the west end and a small library and writing desk at the east end. In the middle of the floor was a large sunken container half filled with liquid. At first, we guessed that this was a toilet but we soon learned that it was simply a water tank.

At the desk sat Yendred's mother, writing something on a strip of material. There appeared to be a light of sorts just over the desk and we were able to trace a wire leaving the back of the light to a nearby battery fixed to the wall. Indeed, there were four other lights in the house, these suspended by wires from the ceiling and powered by individual batteries. To his mother's west were three double bookshelves attached to the ceiling. She straddled a sort of backless chair, or stool, as she wrote.

At the west end of the room was a swing-stair platform and, against the wall, a bed upon which Yendred's father was sleeping. Above the bed was a bottle of compressed gas which Ardeans call "hrabx"—somewhat akin to our oxygen. There are times when Ardean houses must be sealed off. Traditionally, this was a time for the inhabitants to enter a state not unlike suspended animation, as the air would otherwise grow progressively more stale, becoming unbreathable. This continues to be the practice in Vanizla, where the occupants use this time to meditate. The Punizlans,

THE HOUSE

The house was essentially a large, rectangular excavation subdivided by beams which were held together by double spikes and formed into a rigid, well-supported structure. Although it appeared to be very complicated, this structure consisted of only 13 horizontal, 16 diagonal, and 22 vertical beams, hardly enough pieces from which to construct a single, three-dimensional room! It is just as well that the Ardeans enjoy such an economy of means, for their construction methods are equally restricted. Nails are useless since they part any piece of material they may be driven through. Saws are impossible. A beam could only be cut with something like a hammer and chisel. Nevertheless, beams may be attached to each other by double-pointed spikes or by pegs driven into prepared holes. Glue supplements the holding power of spikes and pegs, sometimes even replacing them. In fact, glue comes close to being a universal fastener for the Ardeans, not just in houses but in every sort of construction.

The "swing stairs" are worthy of closer attention. For example, the swing stair at the entrance to the house had a hinge and a spring at one end.

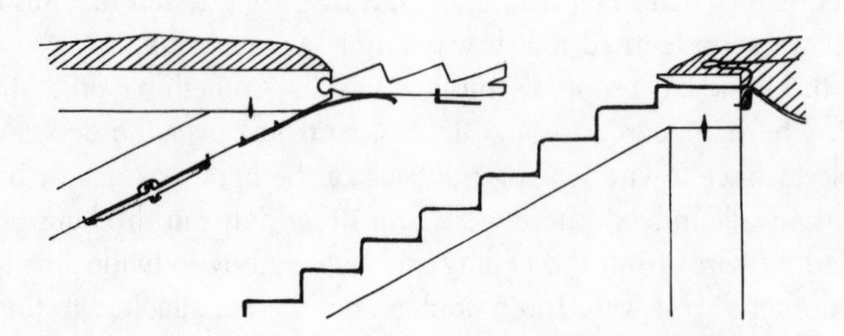

A traveler from the west crosses the entrance to the house by stepping on the swing stair, which under the traveler's weight swings down until it meets the stairway. The traveler descends three steps, ascends three steps, and continues on. Crossing the entrance from the east is only slightly more difficult: the traveler pushes down on the swing stair with his or her lower eastern arm, catching it with the eastern foot and then ascending.

however, merely open the valves of their hrabx bottles and continue with their normal activities.

Each of the beds on the second level consisted of a platform with a soft covering. The Ardeans do not use blankets, since these would interfere with breathing. In fact, Ardeans sleep with one arm over their head, exposing underarm gills to the air.

In many respects, the top floor was the most interesting of all; we watched as Yendred's grandmother trundled back and forth between the stove and the other end of the room. On one of these journeys, she picked up a bowl from a shelf on the east side of the room, transferred it over her head to her upper western arm, and walked with it to the stove, emptying its contents into a large pot in which floated various objects too small to make out at our current magnification.

Ffennell was at the terminal. He seemed fascinated by Ardean food and questioned Yendred closely on the subject.

▪ WHAT IS YOUR GRANDMOTHER DOING?

SHE IS PREPARING FOOD.

▪ WHAT KIND OF FOOD?

SHE ILMA KABOSH WITH BALAT SRAR BOILS. THIS IS MOST TASTY AND WILL IN OUR STOMACHS A LONG TIME STAY. DO YOU ALSO EAT?

▪ YES. WE ALSO EAT. WHAT ARE ILMA KABOSH AND BALAT SRAR?

ILMA KABOSH GROWS ON THE SEA AND BALAT SRAR LIVES AT THE BOTTOM OF THE SEA. WHAT DO YOU EAT?

▪ WE EAT POTATOES AND CARROTS AND SPINACH AND BEEF AND CHICKEN AND MANY OTHER THINGS.

IS THAT THE FOOD OF SPIRITS?

I too was fascinated, but by the cooking process, not the food. Earlier the grandmother had started a fire in the stove by sprinkling something on the pitted surface of the heating unit. Striking this surface with a rod, the grandmother stepped nimbly back from the stove as it gave off a cloud of tiny dots (which we felt safe to interpret as smoke). The smoke made its way lazily up the stairway toward the entrance, soon disappearing entirely over the stove. Some sort of combustion had taken place and the heat generated had been mostly absorbed by the stove itself, heat which now warmed the pot upon the cooking surface. The nature of heat in the Planiverse makes this operation both necessary and possible.

HEAT AND SOLIDS

In our universe, scientists think of the temperature of a substance as the state of activity of its molecules: the faster these move, the hotter the substance. In a solid, the molecules do not move very freely, but vibrate back and forth in more or less fixed locations. As the temperature is raised, the molecules become increasingly energetic and the locations themselves begin to move about, as it were. The substance loses its structural strength. It has begun to melt.

In the Planiverse, exactly the same sort of thing happens but at a much lower level of molecular activity. At a speed which we would regard as not much above absolute zero, most Planiversal solids have already started to melt.

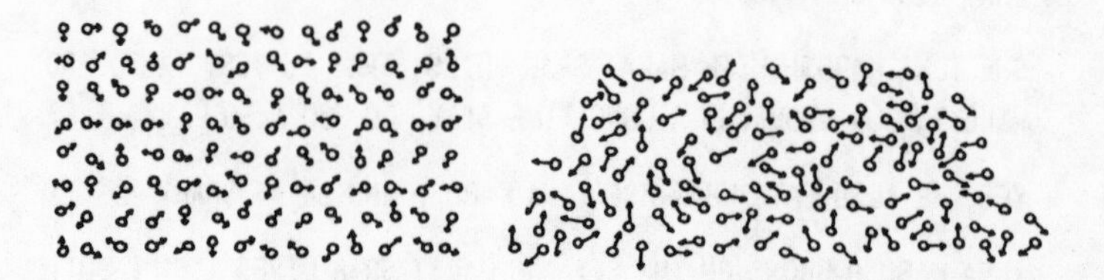

Even before our discovery of the Planiverse, Earth scientists had been examining the behavior of three-dimensional molecules arranged into thin films. Very much the same kind of phenomenon had already been discovered, and even without our contacts with Arde the result could have been predicted.

The chemical reaction used by the grandmother probably generated enough heat to melt common substances on Arde. Yet the material comprising the stovetop was able to absorb much of this heat without melting, releasing it more gradually for the benefit of the stew. Evidently, fires on Arde are extremely dangerous and greatly feared.

Beneath the upper swing stair near the grandmother, we discovered a square device with a very intricate interior. Yendred described its function and we quickly realized that it was a radio. Thinking that we would get around to discussing Ardean radios at greater length later on, we did not bother to find out much about how it worked. (Alas, this was not to be.) The radio was connected by a wire which ran up the entrance stair, doubling as a carpet, to an aerial lying on the ground near the entrance. For receiving hourly information and entertainment broadcasts, aerials are placed in an upright position by plunging one end in the soil. Otherwise, they are left horizontal so as not to inconvenience travelers.

Yendred's sister had stood on the stairway during the first hour of this contact. She shifted her weight slowly and regularly from one foot to the other, occasionally raising and lowering her arms, as though sighing with impatience. She had apparently been forbidden to go down the stairs while her grandmother was cooking, and Yendred had asked her not to interrupt the privacy of his "daydreaming." Through this and numerous other clues we would come to know Ardean society as a very patient and disciplined one, although in this regard the Punizlans seemed to be downright frivolous and indulgent compared to the Vanizlans. Near the end of this contact, Lambert took over at the terminal.

▪ IS YOUR SISTER A FEMALE?

YES.

▪ WHAT IS THE DIFFERENCE BETWEEN MALES AND FEMALES?

FEMALES MAKE EGGS AND MALES DO NOT.

▪ WHAT DO MALES DO WITH FEMALES?

ANYTHING.

▪ I MEAN HOW DO MALES AND FEMALES MAKE BABIES?

THE FEMALE DEPOSITS AN EGG. THE MALE UPON IT SITS. THEN THE FEMALE TAKES IT BACK.

▪ THAT IS TOTALLY WEIRD.

This was a strange thing to say and the other students murmured unhappily. Lambert evidently still lived in 2DWORLD's Astria: Arde was not yet fully real to him.

WHY YOU ''TOTALLY WEIRD'' SAY?

▪ THAT IS VERY DIFFERENT FROM THE WAY EARTH PEOPLE DO IT.

HOW DO EARTH SPIRITS DO IT?

It was pleasant to see Lambert squirm. His face was flushed. Alice watched him intently.

"Go ahead, tell him!"

▪ THEY DO NOT HAVE EGGS.

"Really! Guess again."

▪ I MEAN THEY HAVE ONLY TINY LITTLE EGGS. NOT BIG ONES LIKE ARD [erase] THE NSANA DO.

I will spare Lambert the embarrassment of publishing a fuller account of this conversation. When he finally got out of his dilemma, he obtained many valuable details from Yendred on Ardean reproduction. This information was supplemented on a number of occasions during Yendred's journey across Ajem Kollosh. On these travels we encountered a number of "pregnant" females. Some of these carried an enormous, undifferentiated egg in their lower cavity. This turned out to be not an egg as such, but a large casing containing a womb and related organs. Other females carried a miniature, developing baby Ardean within the casing. Judging from these observations and one scene which we came upon quite by accident, it is possible to infer much about mating and birth: the female develops a large egg in her lower cavity and, when solicited by a male of her choice, is persuaded to release it, casing and all. The male then sits upon the egg and, with feelings that can only be guessed at, fertilizes it with a special organ extruded from his terminal opening. The female then takes the egg back into her cavity (apparently with more pleasure than pain) and the fetus begins development inside the egg.

It is immediately obvious why the Ardeans do not employ a more direct method of fertilization when one considers the Ardean anatomy.

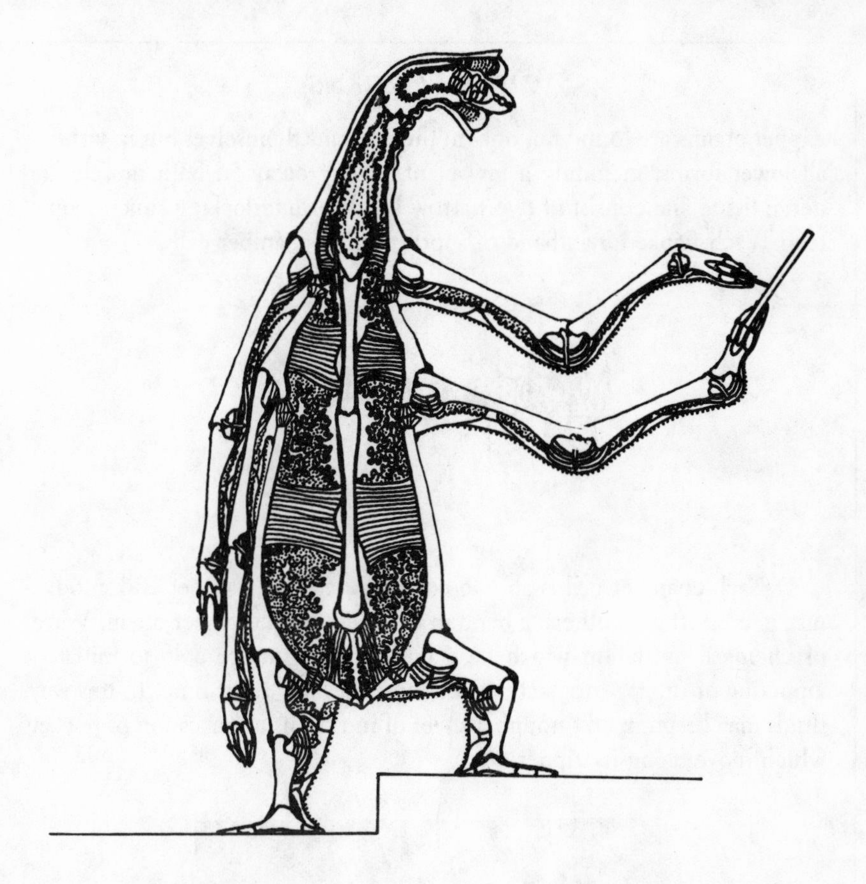

In the Ardean body, we reach the acme of complexity as far as life on Arde is concerned. Structurally, Ardeans have more in common with our insects than with us: their skeletons are on the outside and they have no circulatory system of veins and arteries as we do. Both of these features appear to stem from the same underlying cause: in two dimensions an internal skeleton would severely disrupt the passage of fluids, and tubes of any kind are impossible. Within their exoskeleton circulates a fluid swarming with billions of tiny "carrier cells." The fluid circulates in a clockwise direction around a central muscle platform rather like cartilage in composition. The fluid (which we may as well call "blood") passes through various muscles and nerves by means of a mechanism having no equivalent in life on Earth. We came to call these "zippers."

ZIPPER ORGANS

Zipper organs are found not only in the Ardeans themselves but in virtually all lower forms, including a few plants. They occur in both muscle and nerve tissue and consist of two narrow bands of interlocking (microscopic) teeth, each flanked by a band of short, parallel chamber cells.

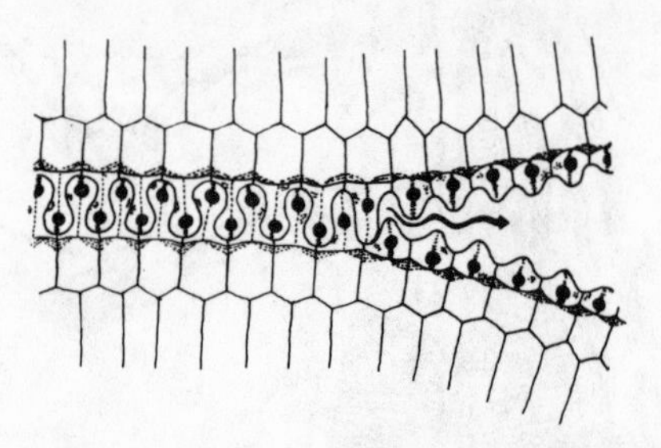

Each chamber cell is able to concentrate fluid at either end, producing, in concert with others, a bend in its section of the zipper organ. Waves of chemical excitation which pass along the organ are able to initiate a zippering or unzippering action in an unbelievably short time. In this way, fluids may be pumped through a sheet of muscle as a succession of pockets which move along its zipper.

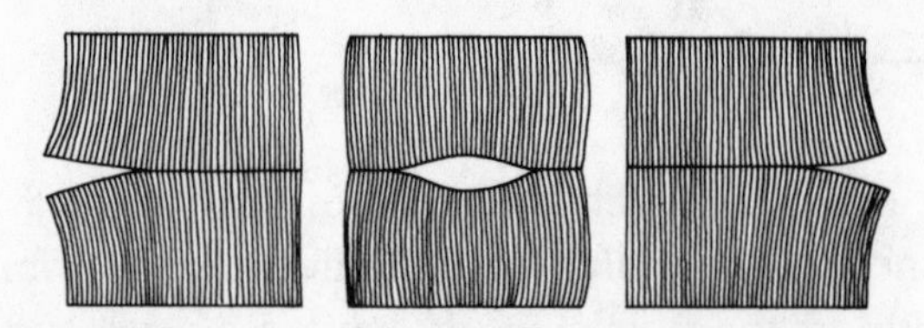

It is even possible for nerves, either isolated or embedded in muscle, to be parted and rejoined precisely without interfering in the zippering process. Naturally, this leads to a degree of intermittent neural action, but all events within the Ardean anatomy appear to be carefully timed, not unlike traffic in a large city.

At the top of the Ardean body is a head containing sense organs but no brain. Two eyes and two ears flank a pair of central, beaklike jaws. The jaws are equipped with small, grinding teeth and are operated by a pair of strong muscles separated by a zipper. The jaw zipper opens directly into a kind of digestive pouch rather like a stomach. Unlike our stomachs, the Ardean stomach has no outlet. Food taken into the stomach is broken down chemically as far as possible and the digestible portion then passes through the wall of the stomach to be absorbed by carrier cells on the other side. The indigestible fraction is then ejected from the stomach by the operation of the jaw zipper muscle operating in reverse. In brief, Ardeans must sooner or later spit up after every meal. Not during this contact but in a later one, Yendred was shocked to learn that we had a "food channel" running through our bodies. His remark at the time was:

WHY DO YOU NOT THEN FALL INTO TWO PIECES?

Ardean blood enters the cavity behind the stomach through one nerve-trunk zipper and leaves the cavity through the other. The nerve trunks connect eyes, ears, taste, and touch organs to the brain, which is held in a cartilage cup at the base of the neck. This entire portion of the Ardean anatomy is held together not so much by zipper muscles as by the careful regulation of internal fluid pressure. This fluid, like water here on Earth, cannot be expanded or compressed.

The Ardean body consists of three segments, each with an appendage; the top two segments have arms attached to them and the bottom segment has legs. Fluid is being continually pumped through the portal muscles adjacent to each appendage. First it is pumped into the appendage, inflating it and bringing in carrier cells with dissolved food. These release the food to tissues and/or absorb dissolved gas at the gills which line the lower edge of each appendage. Then the fluid is pumped out of the limb and into a special cavity just inside the portal muscle. Here, the opposite process takes place: gas is released and food is absorbed by the carrier cells.

Each segment is held together by a muscular sheet which connects the exterior bone to the interior cartilage. Here, the zipper is not in the middle of the muscle but at one side, next to a nerve trunk which follows the contour of the cartilage.

Throughout all the body cavities one finds food storage organs, glands, and other tissues of uncertain function. In the lowermost cavity are found reproductive organs and an opening for liquid wastes.

This brief overview of Ardean anatomy must satisfy the reader for now. Indeed, we ourselves knew much less than this as we examined Yendred and his family. The reader may easily imagine our impatience to know more about this marvelous biological machine, especially the seat of Ardean intelligence—the massive brain sitting on its pedestal of cartilage and conducting its muscular, neural, and biochemical orchestra.

Our discussion of Ardean anatomy with Yendred was interrupted at one point. There was a long delay at our printer while Yendred held his head to the east. Then he "spoke."

A RIVER IS COMING.

▪ HOW CAN YOU TELL?

I CAN HEAR A TRICKLING SOUND IN THE DISTANCE. SOON IT WILL BE HERE.

▪ IS IT DANGEROUS?

I MUST GO INSIDE AND CLOSE THE ENTRANCE.

▪ WILL YOU TALK WITH US?

YES.

▪ WON'T THE OTHERS HEAR YOU?

IT STRANGE IS BUT I DO NOT TALK IN SOUNDS WITH YOU. IT IS IN MY HEAD THAT YOU SPEAK AND THAT I ANSWER. DO YOU TALK IN THAT WAY?

▪ NO. WE TYPE OUR WORDS.

WHAT IS ''TYPE''? WAIT. I MUST GO IN.

With this, Yendred walked to the east with an easy, rocking motion, descended the stairway six steps, and removed a hatch cover held to the entranceway ceiling by two brackets. Ascending a few steps, he inserted the cover into two notches, one in the top entrance step and the other in the bottom step of the swing stair, just above the mailbox.

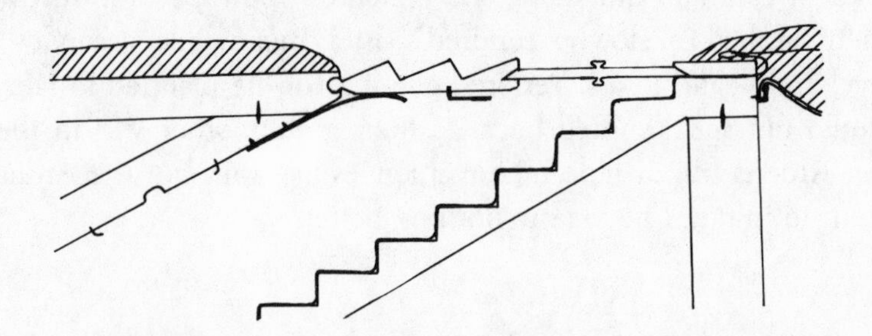

With the entrance thus secured, Yendred made his way slowly down the stair until he came to his sister. The two descended to the swing stair and said something to the grandmother, who stepped away from the stove and down the steps of the cooking platform. First his sister, and then Yendred, used the swing stair. By the time the first layer of running water was passing over the house, the grandmother had walked over to the cupboard so that Yendred's sister could descend to the next floor to turn on the hrabx.

And thus Yendred's family settled in for a long wait as the river passed overhead. This particular river insinuated itself onto our screen as a thin layer covering the soil above Yendred's home and collecting in the swing stair steps before moving on. Gradually the water deepened and the soil and small stones could be seen swirling in pockets of turbulence above the steps. It was the sight of this swirling motion which caused me to imagine the sound it would produce and suddenly I seemed to be a member of Yendred's family, listening to the hollow rattle and rush of the river echoing down the stairway. Should the hatch cover give way, there would be no escape from the torrent of water which would pour down the stairs. We would watch as each level of the house was filled in turn, unable to ascend the stairs to the surface (that rush could not go around us but would force us back with its tons of weight). Covered at last in water, how well would our air-using gills work? Would we struggle and cry out?

Strangely, even as I imagined this Yendred was telling Edwards, who currently manned the terminal, about an identical tragedy which last year had killed an entire family living several houses to the east.

Yendred's father continued to sleep and Yendred's mother had left the desk, making her way to the swing stair near his bed. With every evidence of care and quietness, she pulled on the rope to lower the stair and then climbed up slowly. Yendred's sister, meanwhile, ascended to the top level and began to ask Yendred about a trip he planned to take soon. She could not see Yendred because their grandmother was in the way. Life on Arde is full of little inconveniences like this, but the Ardeans do not seem to mind. They know nothing better.

3

On Fiddib Har

Friday, May 30, 2:00 P.M.

It was nearly two Arde-days before Yendred was to leave on a journey that would span half the continent of Ajem Kollosh. When he came into focus on our screen, Yendred was walking in front of his father toward the shore of Fiddib Har. According to our calculations it was nearly midnight. The two walked without speaking because Yendred was communicating with us much of this time. He had agreed to help his father for the next day. By midnight a strong east wind would be blowing them far out on Fiddib Har, where they would spend the morning fishing. They would then sail back with the afternoon westerly.

As they walked, they passed over the five houses between Yendred's house and the sea. All of these, as well as most of the houses on the other side of Yendred's, were inhabited by fishing families. On Arde, communities tend to form around occupations not only because so much short-distance travel is on foot but because the simple linear geography makes the advantages of living close to one's work glaringly obvious.

Following the progress of Yendred and his father, we became aware that they were part of a procession moving toward the shore. These were fishermen who had all left their homes at the same time. Their boats were stored ahead of them in the same order as their houses, and all would arrive, more or less simultaneously, at their boathouses, no one having to climb over anyone else to get there. Not surprisingly, the boathouses were underground sheds. These lay beneath the sandy kilometer of gently sloping beach between the seashore and the nearest house.

There seemed to be an air of eagerness about the fishermen. I could imagine the brilliant, diamond-studded night sky, its bright glow reflected on the sand, a slowly freshening offshore breeze, and the cheerful banter of beings simplified by the sea. So it was that Yendred and his father arrived at the sixth shed from the water and proceeded to exhume its roof, sliding it toward the shore so that it became continuous with the sloping shed floor. Yendred's father crossed the floor, climbed over the boat, and planted himself behind its stern. Yendred reached into the boat and withdrew a roller disk, which he wedged under the bow. Pushing and pulling, the two rolled the boat up out of the shed and over the roof. His father retrieved the rollers as they appeared from below the stern, tossing them forward to bounce along the horizontal mast to Yendred, who would catch them and plant them once more under the bow. They halted this process only once—to replace the shed roof—and then continued to the shoreline, where the boat slid easily into the calm water.

When the mast had been put up and all the gear and tackle made ready, we realized that the boat had no bow or stern as such. Being completely symmetrical about its mast, bow and stern could be determined only by which direction the boat was moving. This ceased to puzzle us when we realized that a two-dimensional boat could not "turn around."

The mast of the boat has two sections, a stout lower section inserted in a well-braced framework, and a graceful upper section resting on this and held in place by two wedges or pegs. Two long ropes dangled from the top of the mast. There were lockers for food, tackle, and line, as well as two holds for the catch. At either end of the boat were oars. One of these was soon being pumped by Yendred's father; the other was folded up to be used by Yendred on their return. Ingeniously designed, the blades of the oars were hinged so that when the rower pulled down on the handle and the blades stroked toward the rear, they formed a rigid paddle. When the rower pushed up on the handle, the blades moved forward, folding out of the way so as not to counteract the boat's forward motion.

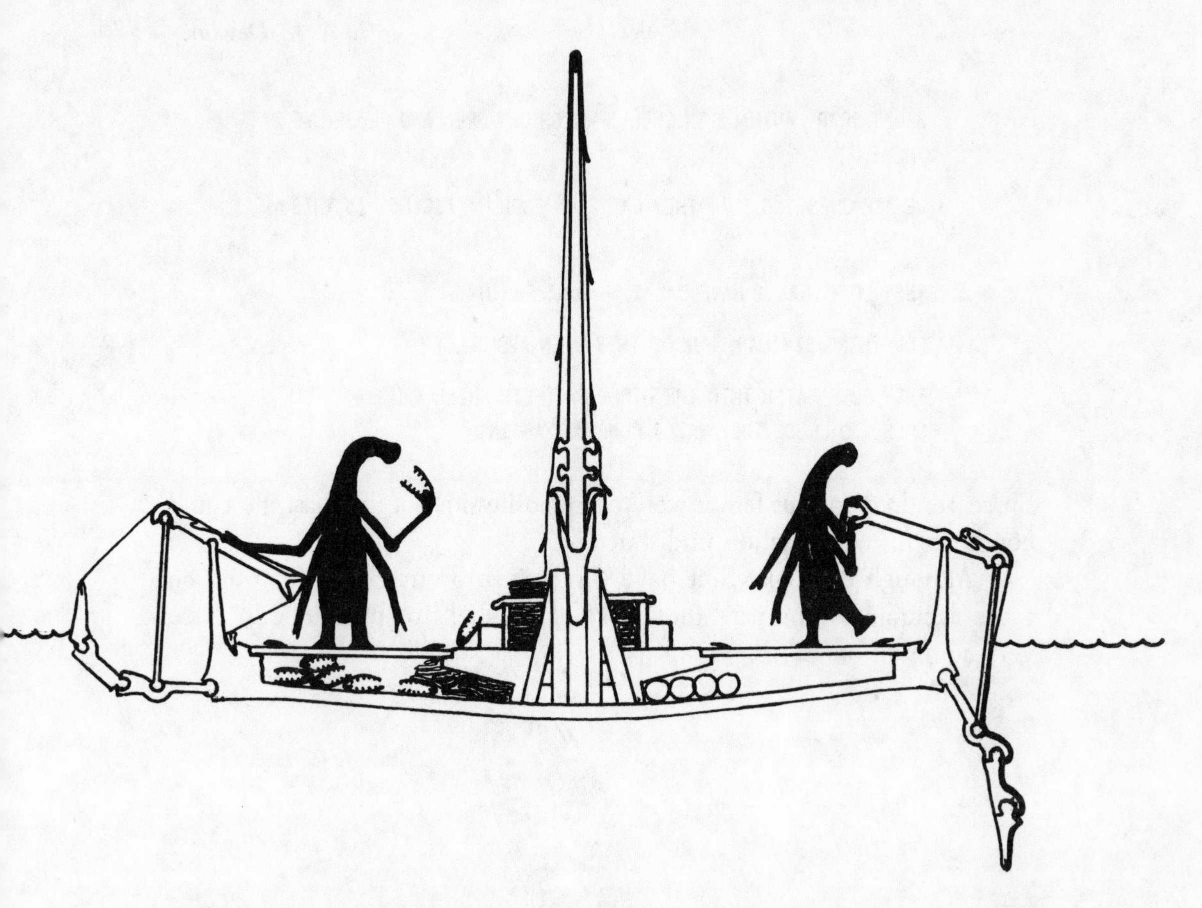

There were no visible fasteners used in the boat's construction; it was all held together by glue.

Although it was night, things appeared on our screen just as they would in the daytime. The sea bottom was barely two meters below the boat and, judging from the boat's progress over colonies of strange shellfish dwelling there, it was now moving forward swiftly. Yendred's father ceased rowing and folded up the oar, securing it with a rope that ran from the tip of the bottom blade up to a peg which he inserted in a slot near his feet. The easterly was clearly strengthening; it blew directly upon the mast, this being all a two-dimensional boat requires by way of a sail.

We asked Yendred about the shellfish which his boat had passed over.

- YOU PASSED OVER SOME ANIMALS OR PLANTS LYING ON THE SEA BOTTOM. WHAT ARE THEY?

BALAT SRAR. ANIMALS IN SHELLS WHICH CAN OPEN AND CLOSE. WE EAT THEM.

▪ WE ARE GUESSING THAT THEY EAT TINY FOOD PARTICLES. IS THIS TRUE?

YES. DO YOU ALSO HAVE BALAT SRAR AT EARTH?

▪ SOMETHING SIMILAR. HOW DO THEY OPEN AND CLOSE?

THEY PUMP WATER INTO THE HINGE OF THEIR SHELL AND FORCE IT OPEN. TO CLOSE THEY WATER PUMP OUT AGAIN.

Since Yendred and his father were on opposite sides of the mast, he could continue this conversation undistracted.

Although the Balat Srar have similarities to our clams, it would be more accurate to compare them with lamp shells or brachiopods, once very common on Earth.

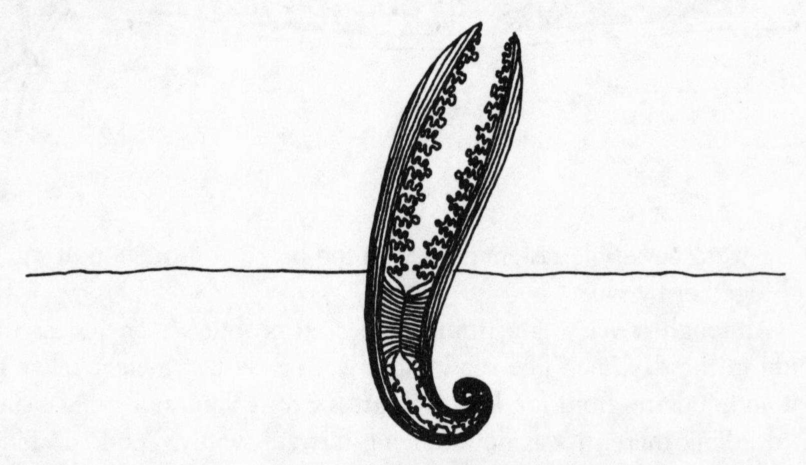

The Balat Srar lives between two curved valves which are hinged at one end. Attached to these valves by a mantle, it consists of fleshy feeding lobes, a "pump muscle," and a pump chamber next to the hinge. A primitive nervous system directs the activities of the Balat Srar. Although clams and lamp shells here on Earth open and close by muscles near their hinge, the Balat Srar does not do so directly. For example, merely contracting its pump muscle has no result since the water in its chamber cannot be compressed. Nevertheless, by employing its zipper organ, it can easily pump water into or out of the chamber, opening and closing its shell.

Watching the Balat Srar opening and closing earlier, and now hearing Yendred's description of their muscular pump, a puzzling feature of Ardean anatomy suddenly became clear: Ardeans move their arms and legs in precisely the same way that this humble shellfish opens and closes its valves. Ardean muscles do not so much contract and expand as they pump fluids into and out of the chamber at each joint. In fact some joints in the arms have two chambers separated by an articulated bone. This gives Ardean arms great flexibility of movement.

Yendred was very enthusiastic on the subject of biology and seemed to know a great deal about the Balat Srar. He explained how the Balat Srar grows by secreting a new layer of shell each year. When fully mature, the male Balat Srar releases sperm into the water at a certain time of the year. Nearby females, which begin rhythmically to open and close their shells, eventually take the sperm into their chambers, lined with unfertilized eggs. When fertilized, the eggs develop into larvae, become detached from the chamber wall, and are expelled into the water of Fiddib Har. At this point, most of the larvae are eaten, but a few survive.

We learned about these larvae much later in Yendred's journey, when we came across the Ardean equivalent of a biological laboratory and had the opportunity to examine the larvae closely and have them explained (through Yendred) by a Punizlan scientist. Apparently, the pump muscle is active from a Balat Srar's earliest moments. It comprises 80 percent of the larva's body mass and, in concert with a smaller, temporary muscle at its head, the larva swims by a kind of jet propulsion, feeding on the organic particles which enter its body cavity in the process.

The young Balat Srar absorbs food into its digestive lobes in exactly the same way as the adult, and the food diffuses through the lobes by a rather interesting mechanism (explained in the Appendix). When the larva reaches a certain point in its development, it ceases to swim, sinks to the sea bottom, and grows two hard shells which spiral as they grow, at first sharply and then more gradually. The spirals meet at one end and form a hinge, which always wears away enough to prevent locking of the valves.

During our discussion of the Balat Srar, the wind was growing steadily stronger and, all too soon, Yendred had to break off in order to tell his father (and us) that they were approaching the boat ahead of them in the fishing fleet. Soon their mast would rob all the wind from the mast of the boat ahead and they would ram it. Should Yendred take out the mast peg or deploy the forward oar in order to slow their craft?

> NOW HE SAYS TO TAKE OUT THE PEG. HE HOLDS THE MAST ROPE ON THE OTHER SIDE. IT IS HARD TO HEAR HIM IN THIS WIND. I HADD USE TO HIT THE PEG. IT IS COMING LOOSE. I PULL IT FROM THE SLOT AND HOLD IT AGAINST THE MAST. MY FATHER LETS THE MAST SWING DOWN OVER ME WITH THE ROPE. HE ASKS IF WE ARE SLOWING. YES BUT WAIT. WAIT. NOW PULL UP THE MAST. HE IS VERY STRONG TO DO THAT IN A WIND. NOW THE MAST UP AGAIN IS AND I HIT THE PEG INTO THE SLOT AGAIN. THE OTHER BOAT IS VERY AHEAD AGAIN.
>
> [Hadd is a kind of Ardean metal.]

In the course of their voyage they had to repeat this maneuver two more times. The wind grew steadily stronger and Yendred's father moved toward the (temporary) stern, occasionally to pull on the rope in order to decrease wind resistance and their resulting speed. The danger was that a sudden gust would cause the mast to break or the boat to pitchpole. As it was, a great mound of water had piled up ahead of the boat as it plowed through the sea. Occasionally, they came upon floating water plants. These would circulate for a while within the bow wave and then slip under the boat. We let Yendred's boat pass off the screen in order to focus upon one of these. It was not difficult to see how it would look when the sea was calm.

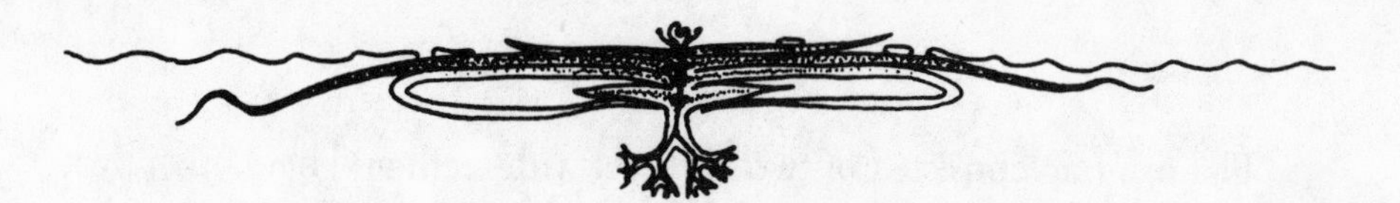

The "Ilma Kabosh," as Yendred called it, has no stalk, merely a sort of root organ joined directly to a symmetrical arrangement of leaves, four on each side. The root organ absorbs minerals and other nutrients from the water through thousands of tiny hairs covering its branched lobes. The bottom pair of leaves produce a gas which fills the buoyancy chambers formed by the second pair of leaves. These curl around and stick to the undersurface of the bottom pair. The third pair of leaves rest upon the buoyancy chambers. These are filled with cells containing "hadrashar," which, according to Yendred, enables the plant to manufacture complex foods from shemlight and the rather simple nutrients it absorbs through its roots. The fourth set of leaves do the same thing except that they can also be moved into an upright position to act as sails. At the very top of the plant is a central arrangement of reproductive organs, a single egg stalk surrounded by two pollinators.

Eventually the wind began to die down and, by early morning, it had become calm enough to fish. Yendred took down the mast, resting it upon the water beside him, and fed line to his father from a locker at the foot of the mast. His father attached some rather vicious-looking traps to the line at intervals and fed these out, over his head, into the water below.

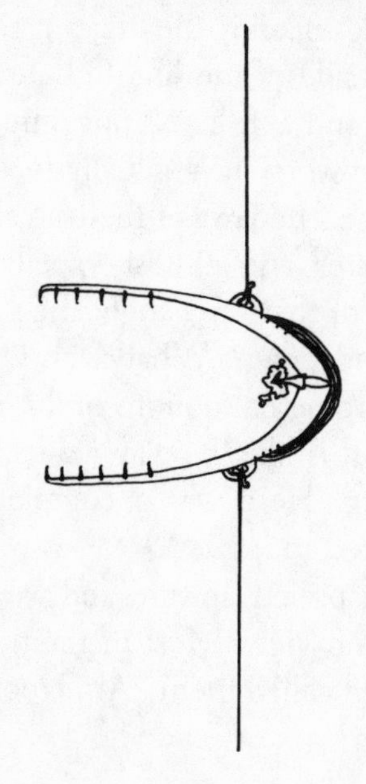

The fish trap consisted of two jaws, set with teeth and held open by a pin placed inside the trap. A morsel of bait stuck to the pin was all that was needed to lure their catch into the trap. As they set the traps and put them overboard, one by one, we tried to imagine what sort of creature they hoped to catch.

The clock on the laboratory wall was nearing midnight our time, even as Yendred and his father enjoyed their midmorning calm. Alice sat at the terminal staring into the screen. Lambert suggested that she follow the line down to see what Ardean swimmer took the bait.

With a sigh, she punched some tracking data into the keyboard and we began to "descend" along the line. Every meter or so there was a tassel attached to the line. Yendred's father had used these to grip the line, as Ardean hands (or *any* two-dimensional hands) cannot grasp a line as ours can.

Alice stopped the scan when she arrived at the first trap: we could not remove ourselves very far from Yendred's current location without threatening the Earth/Arde link. Even as we watched this trap, it began to jerk slightly as the trap below it appeared to have caught something.

"Scan down. Scan down!" Lambert was excited.

"Be patient. I'm sure . . ."

Before she had even finished her sentence, a rather horrible-looking thing swam into view, investigating the trap. Luckily (not for it), the creature was on the proper side of the line to be caught. It looked like a cross between a centipede and a fish. Nosing into the trap, it suddenly seized the bait and began to worry it, gradually loosening the pin. When the pin suddenly popped free, the jaws of the trap closed with frightening swiftness, expelling the water and almost expelling the centipede-fish along with it. But the teeth of the trap caught the creature's fins and held it fast as it struggled up and down. What a curious contrast there was between the beautifully coordinated motions of this organism and the ugly reality which it now faced. It is like the way we feel as children when we see a helpless animal dying. Here was a completely new animal, one whose death I was not inured to.

Abruptly, the trap was pulled upward and we watched as the lower traps glided, one by one, into view. Most of them held fish, which were chiefly of two kinds. Yendred called them "Ara Hoot" and "Kobor Hoot."

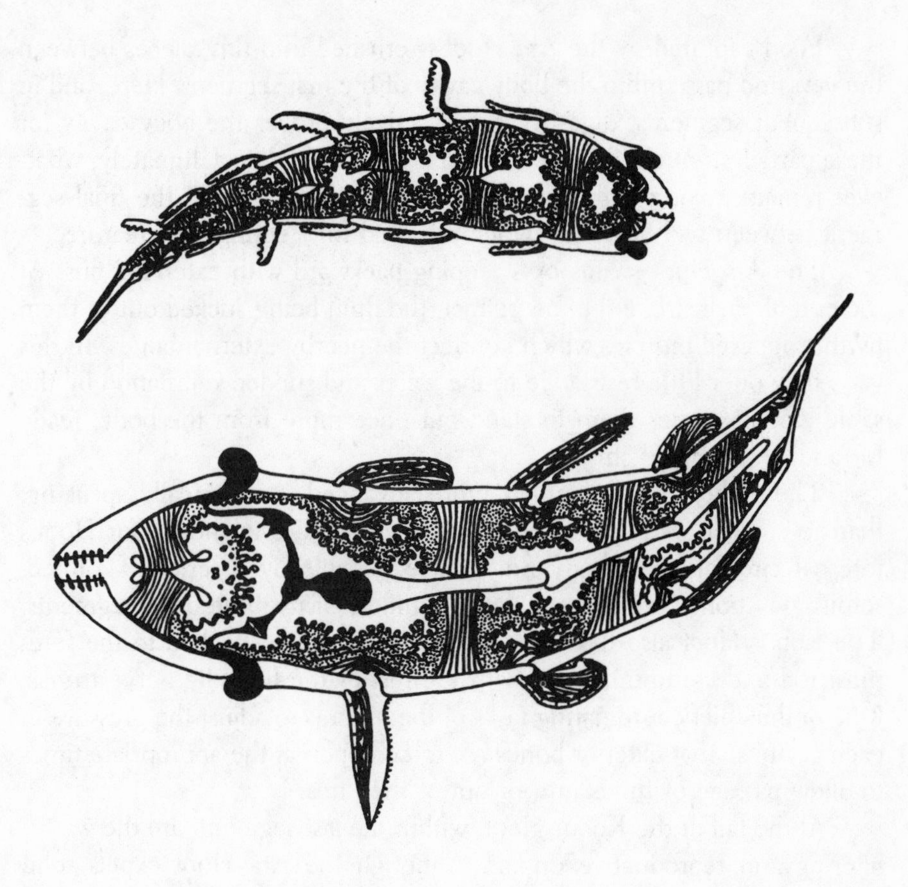

Generally speaking, the Ara Hoot is the smaller of the two fish. It has from four to eight segments, each consisting of a pair of bony outer plates, transverse muscle with a central zipper, and digestive/respiratory tissue. At the head is a pair of hard, transparent eyes upon which the jaws hinge, touch and taste organs, and a very simple brain in two halves. The two halves communicate by a nerve trunk across the trailing edge of the jaw muscle, with the result that every time the creature swallows something this communication is temporarily interrupted. Nerve trunks also run from the brain halves toward the rear of the fish, following the inner surface of the bones as well as penetrating the muscles. Evidently, nerve tissue is not only able to transmit impulses but also has structural strength.

Food captured by the Ara Hoot is crushed into tiny pieces between the jaws and passed into the body cavity of the first segment. Here, and in subsequent segments, digestive tissue actively probes the body cavity for these particles, much like the Balat Srar digestive lobes. Ultimately, whatever remains, along with digestive wastes, is passed out of the final segment between two tailbones which are used for steering the creature.

The Ara Hoot swims by sweeping backward with extended fins. At the end of each stroke the fins retract, the fluid being sucked out of them by the zippered muscles which connect the nearby external bones. In this way, they offer little resistance to the water and sudden reinflation by the same zippers causes them to stand out once more from the body, ready for another rearward stroke.

The Kobor Hoot is much more advanced, anatomically speaking, than its smaller colleague. The chief difference lies in the Kobor Hoot's internal circulation of body fluid made possible by a series of jointed, semi-rigid "bones" running down the middle of its three body segments. The Kobor Hoot also has a larger, integrated brain attached to the foremost rod and communicating with the rest of the body by nerve trunks. One of the chief coordinating tasks of the brain is to adjust the pressure in each cavity so that exterior bones are forced open at the appropriate times to allow passage of fluids into or out of the "fins."

At the tail of the Kobor Hoot, within the last segment, are the waste-filtering and reproductive organs. Although the Ara Hoot expels solid wastes (in particle form), the Kobor Hoot has only liquid waste to get rid of. This is filtered out of the body fluid and periodically released between the tailbones. It is in the last cavity that eggs or sperm are created, and the Kobor Hoot is the only Ardean animal, to our knowledge, capable of direct insemination; the male simply places its two steering bones between the female's and releases sperm. The young Kobor Hoot are nurtured in the tail cavity until developed enough to swim alongside the mother.

As his father pulled in the line and emptied the traps, he pitched the fish across the deck to Yendred, who lifted the cover of the hold and threw them in, one by one. None of them appeared very damaged by the traps and many continued to work their fins, eeling their way forward and back among their fellow creatures within the hold.

SWIMMING AND BREATHING

The swimming fin or "oargan" of the Kobor Hoot consists of a series of jointed bones on its leading edge, a complicated series of muscles next to this, a tissue layer of unknown purpose, a long, narrow cavity, and a gill on the trailing edge.

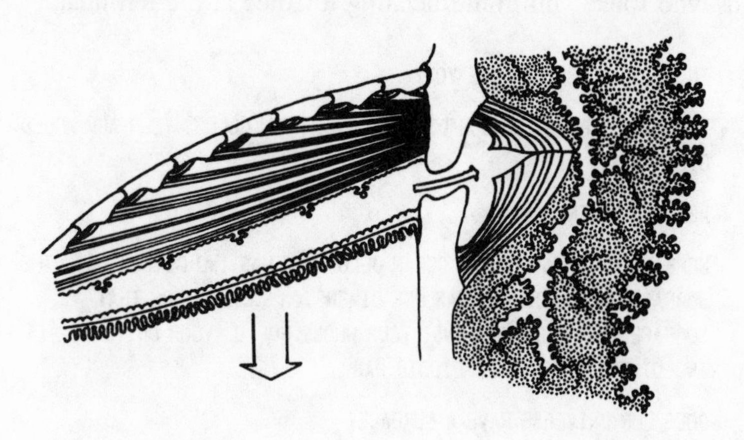

For the Kobor Hoot (and, for that matter, the Ara Hoot), to swim is to breathe. When the fin is inflated, it stands out from the body and the long muscles immediately contract, forcing the fin rapidly toward the rear. At this point in the swimming cycle, the gill at the rear of the fin is most actively collecting hrabx from the surrounding water. Simultaneously, the two body bones are disarticulated and the contracting space inside the fin pushes its fluid through the opening and against the portal muscle. The zipper of this muscle pumps all the fluid into the "pulmonary" cavity even as the bones come once again together and the short muscles of the fin begin their contraction phase. The fin begins to curl as it is drawn forward alongside the body. Meanwhile, having just inflated the pulmonary chamber, the portal muscle now deflates it, creating a bubble against the still articulated bones. When these are forced apart once more, the portal muscle contracts violently, inflating the fin again.

The manner in which nutrients and dissolved gases make their way from cell to cell is fully explained in the Appendix.

The traps were reset and rebaited, and the line was cast back into the sea. The surface of Fiddib Har was absolutely level and calm. Yendred was discussing with his father the long journey he would begin on the morrow. After a time his father fell silent and cocked his head to the east, apparently watching the fishermen in the next boat. Yendred used this opportunity to take up the matter of dimensions with us once again. It was Edwards who spoke with him, dictating to Alice at the terminal.

IN WHICH DIRECTION ARE YOU?

▪ YOU CANNOT POINT IN OUR DIRECTION. IF YOU COULD, YOUR ARM WOULD DISAPPEAR.

WHAT IS DIMENSION?

▪ HERE IS AN EXAMPLE. THE OCEAN BELOW YOU HAS TWO DIMENSIONS. THE SURFACE OF FIDDIB HAR HAS ONE DIMENSION. A POINT ON THAT SURFACE HAS NO DIMENSIONS. YOUR WHOLE WORLD, YOUR UNIVERSE, IS TWO-DIMENSIONAL, LIKE FIDDIB HAR.

DOES OUR UNIVERSE HAVE A SURFACE?

▪ NO. YOUR UNIVERSE HAS NO SURFACE.

THEN HOW IT LIKE FIDDIB HAR IS?

[We paused here, arguing among ourselves how to present this idea.]

▪ WHEN YOU LOOK DOWN AT THE SURFACE OF THE WATER, IMAGINE THAT CREATURES LIKE THREADS LIVED THERE. THEY WOULD NOT GO DOWN, BUT STAY ALWAYS AT THE SURFACE.

IS THAT A WORLD OF ONE DIMENSION?

▪ YES. YOU LOOK AT THAT SPACE FROM OUTSIDE THAT SPACE. THIS IS HOW WE LOOK AT YOU.

Through such conversations, Yendred gained considerable insight into the third dimension, insight which we three-dimensional beings cannot easily appreciate. Although it helped considerably that Yendred had studied rather a lot of science in Punizlan schools, it was clear to us very early on that Yendred, by human standards at least, possessed considerable intelligence. At the same time, we all developed the feeling that intelligence is itself a universal thing, taking one material form or another, but ever striving toward the same goal.

Yendred's father hauled the line in, removed the fish, rebaited the traps, and let them down a third time. The eastern hold was filling with fish and the noon breeze was beginning from the west. We scanned eastward along the fishing fleet and found some of the crews beginning to put up their masts and stow the last catch. When we returned to Yendred's boat, we found Yendred installing the lower mast and his father slowly, almost with reluctance, hauling up the line for the last time.

Now Yendred put their fish, mostly Ara Hoots, in the western hold on his side of the boat. Finally, he took the line from his father and stuffed it all, nearly two hundred meters, into the line box. In doing this there was, of course, no danger that it would tangle. Before long the mast was up and the boat was moving slowly east. Every now and then the boat shuddered as a gust of wind caught its mast. We had noticed this phenomenon frequently during the night voyage as well. It was not until much later, when we could no longer contact Yendred, that we had the leisure to speculate about turbulence in a two-dimensional atmosphere.

Presently, the boat was ploughing eastward through the water, Yendred and his father having exchanged roles. Now it was Yendred's turn to hold the mast rope and spill air whenever it became necessary. Although the winds of Arde always come at the same time and from the same direction with almost perfect regularity, their force can vary considerably. On this occasion, the wind was almost visible as the mast quivered and the boat bounded over the foam. Water was collecting in the forward end of the boat and we feared that it might be swamped until we saw Yendred's father drag his lower, eastern arm along the floor of the boat, trapping all the water ahead of it. Following the inner contour of the boat, he literally swept all the water up the stern, over the oar, and out into the churning sea.

Suddenly, to our horror, the mast snapped just above the two wedges. Yendred had given the rope too much play and now it zipped from between his fingers and the mast struck his father's head, falling to the deck and leaning against his still upright body at a crazy angle.

TURBULENCE

Turbulence in air is a rather complicated flow in many different directions. Often, turbulence is composed of vortices or currents of air following a circular pattern, usually in the wake of some obstacle past which the air is moving. In three dimensions, the turbulence dissolves, so to speak, when its large vortices break up into smaller ones.

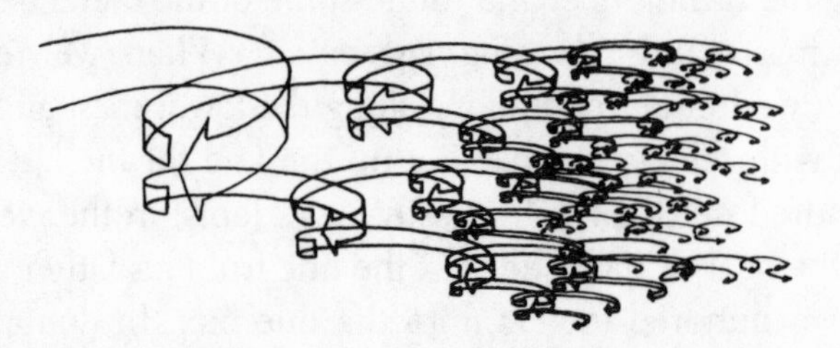

In two dimensions, turbulence does not dissipate nearly so readily. The vortices continue to propagate themselves for considerable distances along the wake of an obstacle.

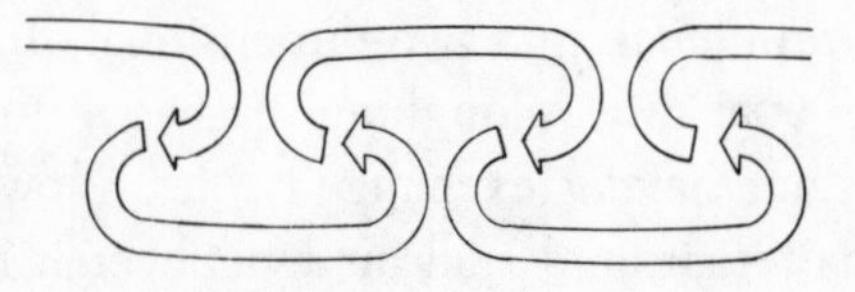

Had Yendred's father been human, the force of the blow would undoubtedly have knocked him unconscious. Luckily, Ardean brains are not in Ardean heads and his father now stood his ground, waiting for Yendred to clamber over the stump of the mast and, balancing in the gale, to jump into his father's side of the boat.

Together they removed the stub of the old mast, pulled out the wedges, and replaced the now shortened mast in its hole. Behind them, to the west, five boats were spilling wind and plying their eastern oars, trying not to ram Yendred's boat or one another.

The six boats, with Yendred's in the lead, now made their way east in the gale at somewhat reduced speed. Yendred held the mast rope as firmly as he could, his two fingers surely aching.

As the sturdy craft bounded over the water of Fiddib Har, we zoomed in on the father's head to examine it closely. There appeared to be no damage either to his eye or jaw. Both were constructed of hard material. But the tissue below the eye on his father's western (left) side seemed to be displaced slightly. Would this not cause him pain? We wondered how Punizlan doctors (if there were any) might repair such damage. Surgery would certainly be a very different affair on Arde than it was on Earth. For one thing, cutting would not be done by a knife but by a needle. For another, suturing was impossible. We were to discover much later on that very little surgery is done on Arde, that most external wounds are treated with clamps and that the greater part of Punizlan medical practice was based on orally administered drugs.

By late afternoon the wind was dying and the last of the fishing fleet arrived at the beach. One by one the boats were rolled up on the firm sand, but they could not be as easily moved as on the previous night. The wind was now onshore and great breakers, the remnant of the afternoon westerly, broke against stumbling Ardean forms. Even so, the fishermen appeared to be disembarking much farther up the beach than where they had earlier set out. Without a moon, Arde could have no lunar tides. Yendred explained that this high water level was due entirely to the pressure of the west wind acting upon Fiddib Har. Even as the two pulled their boat up on the beach, one could see the water slowly recede.

Before they had gone a few meters Yendred and his father encountered an obstacle. A giant carnivorous plant called a Mil Dwili had washed up on the beach.

It is not hard to imagine how this plant must look when floating on the surface of Fiddib Har (see next page). It stays afloat by means of several buoyancy chambers and traps large fish between its two dangling arms. A system of roots then grows directly into the prey, absorbing its organic material. Was this an animal or a plant?

Yendred did not give us a clear answer. Since the boat could not be made to roll over the plant, his father picked it up and placed it under the mast. They rolled the boat up the gentle beach towards the shed, Yendred throwing the rollers to his father as they emerged from beneath the boat. Stowing the boat in the shed, they removed the day's catch and divided it up into two string bags. Then they replaced the shed roof, leaving the Mil Dwili to rot in the sand.

This was to be the longest period of continuous contact with Yendred we would ever experience. For almost 24 hours we had followed the fishing expedition. Outside the laboratory window burned the early afternoon sun. We were exhausted.

And so were Yendred and his father. They walked wearily home, each carrying a bag of fish. His father complained once about the soreness in his eye but said nothing else to Yendred. Yendred said nothing to us. Soon the sun would set in the west.

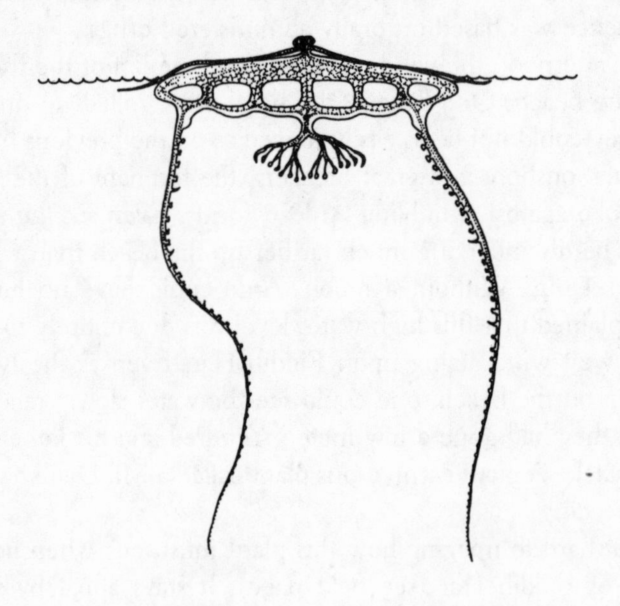

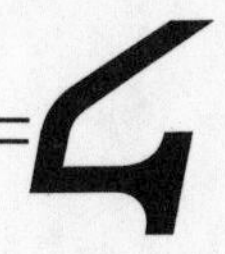

Walking to Is Felblt

Sunday June 1, 2:00 P.M.

Toward the end of the Fiddib Har contact, Yendred had agreed with us on a time for the next contact to begin. He had said that he would be walking to Is Felblt, the capital city of Punizla some 20 kilometers distant. When the contact was renewed at the agreed-upon time, we found him still in his house. The leavetaking procedure had apparently been prolonged.

He was standing beside his mother, who first handed him some money consisting of folded strips of some material, and then a letter addressed to one of her relatives. Yendred placed the money and letter in a special traveling pocket which he had earlier glued to one of his bones.

His mother then did a rather peculiar thing: she ran a finger along the undersurface of the two arms on the side of Yendred nearest her. She was apparently feeling his gills, but the gesture had a ceremonial quality about it.

- WHAT IS YOUR MOTHER DOING?

SHE IS CONFIRMING MY HEALTH BY TESTING FOR WETNESS ON MY GILLS.
SHE WISHES CONTINUED HEALTH.

▪ ARE DRY GILLS A SIGN OF SICKNESS?

YES. WITH DRY GILLS WE CANNOT BREATHE AND WE DIE.

During this portion of the conversation, Alice, who had been taking notes to my left as I sat at the keyboard, stopped writing suddenly and picked up our 2DWORLD manual, thumbing hastily through its pages.

"Look," she whispered. "Here."

Distracted, I read the title of the section she had found: "Eavesdrop Facility."

"Do you think it might work?" I asked.

"Why not?" Her eyes were bright, filled with a curious confidence, as though she already knew it would.

▪ EAVESDROP

YOU MY EGG AND SON ARE. WHY YOU HOME LEAVE MUST? WHAT THINGS BEYOND ARE? HERE YOU EVERYTHING NEED IS. IF YOU GO MUST. MY BLOOD WITH YOU GOES.

I YOUR EGG AND SON AM. THINGS THERE ARE WHICH NOT HERE ARE. WHICH THINGS I NOT KNOW. YOU IN MY EYES WILL BE.

This bizarre conversation took some sorting out at the end of the contact. First, it was obvious that the speech patterns Yendred had been using with us were not his own but had been learned during the initial few days of our contact in May. Second, we concluded that "My blood with you goes" was probably some sort of farewell formula expressing heartfelt wishes—except Ardeans have no heart in the usual sense of the word (their entire body forming one large heart, as it were).

We continued eavesdropping as Yendred went up the stairs to the surface, followed by his father. The rest of the family had been asked to stay behind. We discovered the reason for this when Yendred reached the surface, crossed the swing stair, and waited for his father to emerge. They talked:

I THIS BEFORE HAVE SAID NOT. YOU NO BROTHER HAVE AND I BUT ONE SON. CARE HAVE AND TO US RETURN. MY BOAT YOURS WILL BE. STRANGE

IT IS BUT SOME TO YEARN BEGIN AND LEAVE MUST. THEY TO VANIZLA GO AND THEN RETURN. THEY NOTHING FIND. OTHERS RETURN NOT.

I OF THIS KNEW NOT.

IT RARE IS BUT I OF THEM HAVE HEARD. OF THOSE WHO RETURN BE AND CARE HAVE. MY BLOOD WITH YOU GOES.

YOU IN MY EARS WILL BE.

With this, Yendred disappeared off the right side of our screen heading east toward the city of Is Felblt and leaving his father, who watched him go with both eyes, one not seeing very well.

We caught up with Yendred as he passed over the doorway of the house just east of his own. At first we were eager to confirm our ability to eavesdrop. But then I thought the better of informing Yendred: it might inhibit him in dealings with fellow Ardeans. There was yet another sense in which we might unduly influence events on Arde:

▪ YNDRD WE MUST ASK A GREAT FAVOR FROM YOU.

HOW CAN I POSSIBLY DO A FAVOR FOR THE EARTH FRIENDS?

▪ PROMISE THAT YOU WILL NEVER TELL ANY OF THE NSANA OF OUR EXISTENCE.

THAT IS EASY. THEY WOULD NOT ME BELIEVE IF I DID.

For the next ten minutes Yendred walked with a relaxed, unhurried oscillation, a pace designed not for speed but for distance. We talked about houses, those who lived in them, and their relation to the Ardean surface. Apparently the ground of Arde, even in progressive Punizla, has a sacred, inviolate character. One may possess one's house and everything in it, but the surface is no one's property and there are very strict rules about how it must be treated. No Ardean of either nation may lay claim to ownership to any portion of it—indeed, such a prohibition is so deeply rooted in the Ardean psyche that it rarely occurs to anyone to make such claims. Moreover, since plants, burrows, and other evidence of life interrupt the surface, none of these may be disturbed either. Neither is any littering permitted.

Because we were tracking Yendred so closely, he saw the approaching Ardean before we did. He alerted us that someone was coming and, presently, another Ardean entered the screen from the east. It was a female, judging from her enlarged womb area. She was clearly going in a direction opposite Yendred's. This was the sort of impasse we had been anticipating: one of them was going to have to walk over the other.

TSSIL-GREETINGS. [Said Yendred.]

YNDRD-GREETINGS. WHERE YOU DO GO?

THIS MY JOURNEY TO VANIZLA IS. I FOR A LONG TIME WILL BE GONE. PERHAPS FOREVER.

FOREVER? THIS TO LEARN THE KNOWLEDGE OF BEYOND IS? WHY THAT FOREVER SHOULD TAKE?

I TO COME BACK HOPE AND I YOU AGAIN TO SEE HOPE AND THIS TO SHOW THAT HOPE IS.

Here Yendred raised his upper arm, but Tssil said

WAIT. FIRST OVER ME PASS.

In Punizla, it is the practice that those traveling east have the right to walk over those traveling west. Perhaps the theory behind this is that eastward walking requires slightly more effort as it is always on an uphill grade, however gentle. In Vanizla, the rules are far more complicated: essentially they blend the two principles that physically weaker citizens pass over physically stronger ones and those of higher station pass over those more lowly.

Perhaps it was in anticipation of the Vanizlan practice that, as Tssil prepared to lie down, Yendred suddenly fell prone before her. Tssil hesitated and then stepped forward onto Yendred's neck, then his arms and leg. With a delicate hop, she was once again on the ground and Yendred righted himself.

▪ YNDRD. LAMBERT HERE. WAS THAT PAINFUL?

NOT AT ALL. BEING WALKED UPON ONLY MEANS THAT ONE CANNOT BREATHE FOR A SHORT TIME.

▪ WHY DON'T YOU PUT ONE ARM OVER YOUR HEAD WHEN SOMEONE WALKS OVER YOU?

THAT WOULD BE PAINFUL. OUR GILLS HAVE MUCH FEELING.

As he said this, Yendred raised his upper arm nearest Tssil and she did the same. They then briefly touched their gills together.

▪ WHAT ARE YOU DOING?

Yendred did not reply until he had taken his final leave of Tssil. He explained that this was a form of parting between males and females having a certain relationship. Did Earth people touch their gills together?

▪ WE HAVE NO GILLS. WE TOUCH OUR LIPS. IT IS CALLED ''KISSING.''

WHAT ARE ''LIPS''?

▪ LIPS ARE SOFT PARTS ON OUR MOUTHS.

THAT SEEMS STRANGE. PERHAPS IT IS ''TOTALLY WEIRD.''

We pressed Lambert to probe Yendred about the emotional rather than the physical side of his relationship with Tssil, and we presently learned that in a fit of passion he had once asked for her egg but she would not release it and then, some time later, she had offered her egg but he had refused it. Both had retreated in the knowledge that acceptance of the egg was tantamount to marriage. Apparently Yendred already had his quest in mind.

For the next several hours Yendred trundled over the monotonous landscape of Arde, passing the occasional burrow or home entrance. He trod gingerly on the leaves of flat-lying plants when he encountered them.

All the vegetable forms of Arde have this same low profile. Typical of these, and the most common plant encountered by Yendred, is the Jirri Basla. It consists of two long leaves which lie flat upon the ground on either side of a shallow, but well developed, root system. At the center of the plant is a set of specially modified leaves wrapped around each other to form a bulb, a food-storage organ which, when it is ripe, may be harvested by Ardean farmers (and only by farmers).

When a river comes, the Jirri Basla is able to detach itself from its roots and float: the bulb contains a few layers of trapped air which ensure its buoyancy. When the river peters out, the Jirri Basla finds itself settling upon wet, fertile soil. A new root system develops within hours and the plant takes up existence once again as though nothing had happened. Few plants, whether they germinate on the slopes of Dahl Radam or elsewhere on the continent, travel thus more than two or three hundred kilometers during their lifetimes.

When the bulb of a Jirri Basla is removed, a set of reproductive organs quickly develops from the bulb scar. Two seeds result when a pair of ovaries are fertilized, either by the plant's own pollen or that of a neighbor.

The resulting seeds are each equipped with five flight hairs. In Punizla, these open up at noon so that the afternoon westerly may carry them toward Dahl Radam. By sunset they have fallen to the ground and then they close up again. A species of Jirri Basla also inhabits Vanizla, identical in all respects to its Punizlan counterpart except that its seeds open at midnight and close at dawn. When either kind of seed reaches the height of Dahl Radam, the increased altitude apparently triggers a change. The flight hairs are shed and the seed germinates within the first wet soil it encounters.

All of this was related to us by Yendred, who presently came upon a rather dense patch of Jirri Basla.

THEY IN SHEMLIGHT SHIMMER
THE UNMOVING RIVER
THAT MY EYE IN GREEN DROWNS
AND IN BEAUTY COVERED LEAVES.

▪ IS THAT A POEM?

IT WAS WRITTEN LONG AGO. IT IS ABOUT THE JIRRI BASLA.

▪ WHAT COLOR IS GREEN?

GREEN IS THE COLOR OF JIRRI BASLA. [Yendred paused.]
GREEN IS GREEN.

The poem referred to the appearance which a "field" of Jirri Basla has to an Ardean. The shimmering is caused by the slow and regular way in which each plant places one leaf or the other on top of its neighbor's in order to receive a greater share of shemlight. In a short time, however, its own leaves once again find themselves in the shadow of adjacent plants. The resulting slow-motion competition for light not only ensures that all plants receive an equal share but creates a shimmering effect as the translucent upper surfaces of their leaves catch the light.

This interesting reference to how Ardeans see things triggered a long exchange with Yendred about how he viewed the scene before him. Although we three-dimensional beings could witness Yendred's world "from the side," as it were, seeing everything (inside and out) with perfect clarity, it was very hard for us to imagine how Ardeans saw things. It is paradoxical and yet, perhaps, only fair that seeing so much more of Arde

THE EYE

The Ardean eye consists of a strip of roughly 10,000 facets arranged along a semicircular surface. Each facet caps a long, thin sensory cell and these form a second layer behind the facets. Each sensory cell gives rise to a nerve fiber which reduces or refines the information to be sent down to the brain: for every eleven or twelve sensory cells, only one nerve fiber leaves the eye.

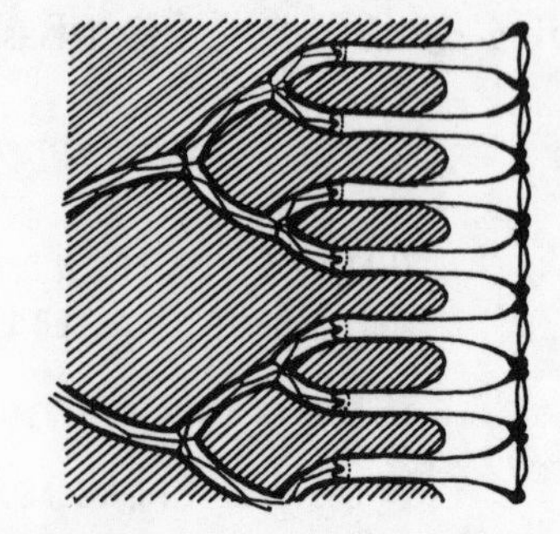

This diagram is largely conjectural. Individual visual cells were barely resolvable by our system, and the precise details of nerve interconnections were nearly impossible to make out. Presumably much of the shaded area is packed with supporting cells which transport energy and nutrients to the nerve and sensory tissue.

Each facet is responsible for covering a very small angle of view within the total visual field. By means not yet understood by the Ardean scientists themselves, all of these small pieces of visual information are incorporated into a vast, parallel network of something like a thousand nerves, which leave the eye in a single trunk.

from our point of view made it that much harder to see it from the Ardeans' point of view or to appreciate their visual modality.

The following example affords a small insight into the problem. Imagine an object the shape of a five-pointed star within the Planiverse. From without we see it as we normally do. In order to appreciate how an Ardean might see it, however, we must turn the star sideways.

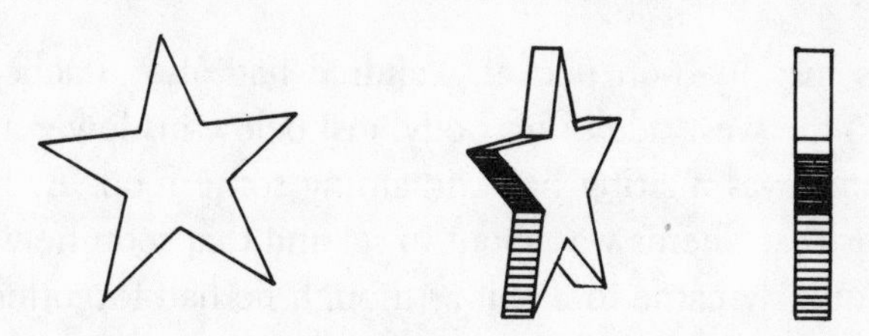

Naturally, being two-dimensional, the star disappears from view when seen edge-on by three-dimensional beings like us. I have therefore artificially thickened it to convey some sense of the star's appearance to the Ardeans. Everything they see is like this: their entire visual world is organized into a single line which encircles them. The line is subdivided by colors and shades, by dark and by light. I have tried to imagine myself as an Ardean looking out at such a linear world. I have even imagined the line with thickness but found that I could not easily tell what I was looking at. Only Ardeans can do that.

On and on Yendred plodded. In our laboratory it was getting late in the evening, and yet it was still afternoon in Punizla when Yendred passed an underground structure which turned out to be a kind of local depot in the Punizlan distribution network.

From such depots Punizlans may buy what they like, although only those living relatively close to the depot use it as we would a store. The rest are content with a delivery service conducted by a few employees of the depot who carry the day's orders east and west to local residents. It puzzled us how anything short of a small army could handle this distribution, since several hundred homes were served by each depot. It was then that we first learned of the Punizlan ability to construct lighter-than-air balloons. Those charged with delivery of goods make up a large package and enclose it all in a rope, then attach the rope to a balloon and, guiding the balloon by another rope, walk easily—or even leap—over the surface from house to house. Several days later we were to see one of these.

Yendred passed over the depot. Just east of the entranceway were a number of clamps, anchored in underground blocks and barely protruding from the soil. These, Yendred explained, were mooring blocks for the much larger freight balloons which made stops at the depot every eight days.

Besides his glued-on pocket, Yendred had also attached an all-purpose catch to the west side of his body, just below his lower arm. Hanging from this catch was a string bag containing some food. As he walked he was telling us that Shems was about to set and that soon he would have to eat. But he quickly came to a halt as though he had forgotten something.

▪ WHY DID YOU STOP?

THE GROUND DOES NOT FEEL GOOD. I AM SINKING.

Indeed, he was sinking. Very slowly, his feet were being engulfed by soil. This was one of those treacherous hollows we had heard about.

▪ WHY DON'T YOU WALK BACK WEST AGAIN?

PERHAPS I WILL NOT SINK FAR.

"Good God, man," breathed Alice. "Get out of there!"

We should have known that Yendred's experience with such soft spots would tell him almost immediately whether they were dangerous. At this point I had the impression that Yendred was deliberately being dramatic. All the same, I made a mental note that this soft spot in the soil did look slightly more watery than the ground Yendred had been walking over earlier. I instructed the students to watch for these so that we might be of some service to Yendred on future occasions.

He lifted his feet out of the mud and plopped them down again, now traveling at only half his earlier speed. Yendred took this obstacle rather philosophically and spoke of soil and water as he went. Although the water cannot leave the soil by drying out, there are other mechanisms for doing this.

THE WATER-SOIL CYCLE

During the winter rivers are quite frequent on Arde, forming almost daily everywhere on Ajem Kollosh. During this time the amount of sediment thus transported from Dahl Radam increases and a gradually thickening blanket of wet soil is laid down over Punizla and Vanizla.

The soil cannot dry out by itself, but when the summer arrives and rains are less frequent, it becomes dry in two ways. First of all, numerous small burrowing creatures, such as the Tahti Neml, move through the wet soil at all times. Now, during the summer, whenever they come to the surface all the soil on the edge of their burrows (whatever portion has not collapsed) may dry out.

A second mechanism for drying the soil is created by the abandoned root systems of Ardean plants. Small organisms akin to our protozoans eat their way up and down the rotting roots, which then dry out, shrink, and expose the soil they formerly penetrated.

The combined effect of millions of such burrows and cracks in the soil is enough to render it almost completely dry by the end of summer, so much so that great dust storms sometimes blow across the face of Punizla and Vanizla at that time.

By 9:48 P. M. our time, Shems had set for Yendred. The soil he now walked upon was reasonably firm and he decided to stop for his meal of the day. The anatomy of Ardeans does not permit them to "sit down" as we do, but Yendred did the equivalent thing by shuffling out two holes for his feet. Digging with quick, flipping motions, he soon had dug himself into the ground to the point where his bottom was only centimeters above the soil. Then the muscles of his legs visibly relaxed and he settled a little. Strangely, his body still did not touch the ground as we thought it would. It was Chan who finally realized what was happening, pointing out that all the air between Yendred's legs was trapped between his body and the ground. He had simply provided himself with an air cushion, his legs locked in place by the two pits he had dug.

All afternoon, of course, the westerly wind had been blowing on Yendred's "back," helping him somewhat on his journey. But now, after sunset, the air was perfectly calm. He sat upon his air cushion and alternated between placing fragments of food between his jaws, head to one side, and then looking upward.

▪ WHAT ARE YOU LOOKING AT?

THE STARS ARE ABOVE ME. THEY ARE A BEAUTIFUL VISION THROUGH THE CLOUDS. FROM THE EAST TO THE WEST THEY LIE ON A GREAT CIRCLE. DO YOU ALSO HAVE STARS?

▪ YES. AND OUR STARS ARE ALSO BEAUTIFUL. THEY LIE ON A GREAT SPHERE.

WHAT IS A SPHERE?

▪ IT IS HARD TO EXPLAIN. WE COULD SAY THAT A SPHERE IS A CIRCLE OF CIRCLES.

THAT IS A BEAUTIFUL SAYING FOR A DIFFICULT IDEA. I SHALL REMEMBER IT.

This idyllic scene lulled us all into a sense of calm and relaxation. The computer hummed gently in the background, lights playing lazily across the multiplier-quotient display. Alice watched the screen contentedly while Chan made notes about the air cushion idea.

▪ ARE YOU GOING TO WALK ALL NIGHT?

NO. IT WILL RAIN SOON AND I WILL TAKE SHELTER WITH A FAMILY.

A gentle rain often falls early in the Ardean night, due to the condensation of water vapor as the Ardean atmosphere cools after sunset.

Within the hour it began to rain, hardly enough to form a river but enough to cause Yendred some discomfort. This discomfort was not due to the wetness of the rain but its chill. Ardeans do not normally wear clothing of any sort. This is partly because the Ardean climate is so equable and partly because the thick and rigid Ardean exoskeleton protects them from the elements. In any event, clothes could only consist of strips of material taped or glued in place. To fend off the rain's chill, Yendred took out a length of string from his bag. He would walk for a time holding the string over his head, but could not see anything while doing this. Thus every minute or so, he would take away the string in order to look in the direction of his travel, watching for the telltale swing stairs which would betray the presence of a snug house.

Presently he came to a house. Its hatch was in place and Yendred tapped upon it with one of his toe bones. We scanned downward into the house and saw an Ardean leave his bed on the second floor and make his way upstairs.

▪ SOMEONE IS COMING. YOU ARE IN LUCK.

IT IS NOT LUCK. THEY MUST COME. IT IS OUR LAW.

The Ardean removed the hatch, getting a little wet in the process, and bade Yendred enter. Being the last down the entrance stairs, it was Yendred who replaced the hatch.

The two made their way through a kitchen and down a swing stair, where his host showed Yendred into a small alcove with a bed. He then proceeded down to his own sleeping quarters after expressing very briefly the wish that Yendred have a restful night. Ardeans have only one sleeping position, on their side. Two arms forming a cradle for his body, Yendred lay for a time musing "out loud" about the wisdom of his journey. What did we think? I thought for a time and then instructed Alice to type:

▪ EVERY INTELLIGENT BEING MUST SOONER OR LATER EXPLORE ITS OPTIONS.

It was midnight and a good time for us to go to bed as well.

Sunday, June 1, 2:00 P. M.

Luckily, the Ardean sleep period is nearly two hours longer than that of a human, their day being nearly thirty-two of our hours. Three of us were back in the lab shortly after the Ardean sunrise when Yendred awoke. We sometimes had to pretend to be doing other things when students and faculty were about. Anyone wandering in out of idle curiosity was delayed by some artifice at the door, giving us time to warn Yendred, terminate the 2DWORLD program, and run an innocent-looking game of Space War in its place. How dreary to launch photon torpedoes at an attacking Klingon warship knowing that Yendred stood waiting on Arde, wherever that was.

As chance would have it, no one interfered on this particular occasion and Yendred, after a quick breakfast with his host and family, departed up the entrance stairs and emerged onto the surface once again into the morning calm. He set off in the direction of Is Felblt, the Punizlan capital, now less than a day's journey to the east.

This portion of Yendred's trip was unremarkable. On the previous day he had crossed the occasional swing-stair entrance as he made his way toward Is Felblt, but today these entrances were much more frequent: he seemed to cross one every forty meters or so, a sign, perhaps, that he was nearing the city. By afternoon the wind was once more blowing strongly from the west.

I have already mentioned that Ardean weather is very regular and predictable, at least in comparison to our own. Every day follows the same cycle: morning calm, afternoon westerly, evening calm, pre-dawn easterly. The explanation for these calms and winds lies with two atmospheric regions, one of high pressure and one of low pressure, which daily circle Arde, always keeping pace with Shems.

The region of high pressure is caused by the expansion of the Ardean atmosphere on the warm, daylight side of the planet and by its contraction on the night side, where it is cooler. Winds circle these regions, in a sense, and give rise to the afternoon westerly and the pre-dawn easterly.

In the late afternoon, Yendred reached the outskirts of Is Felblt and here he encountered a whole group of fellow Punizlans playing a strange

WEATHER PATTERNS

Punizlan scientists have observed that air pressure is normal during the pre-noon and pre-midnight calm periods. But during the afternoon the pressure is higher than normal, and before dawn it is lower. Piecing together their observations and theories, we have arrived at a schematic diagram which summarizes everything very nicely and is quite consistent with our own meteorological science.

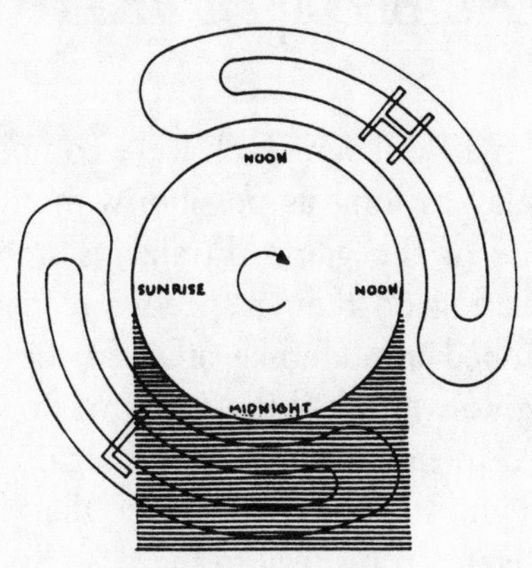

The region of high pressure extends from mid-morning to just after sunset, while the low-pressure region covers the complementary portion of the planet's surface. These regions are stationary with respect to Shems, Arde rotating beneath them, as it were.

On the other hand, Ardean air rotates with the planet, the surface air speeding up under the high (observed as a west wind) and slowing down under the low (observed as an east wind). The relative circulation of air about the high and low is caused by coriolis forces, the deflection of air in an expanding region into a clockwise flow, the opposite deflection operating for a contracting region. The coriolis force, a well-known phenomenon on Earth, is itself simply the tendency of any particle in a rotating field to "slip" in a direction at right angles to its natural motion, like a ball rolling across a phonograph record.

The region of high pressure is caused by the air on Arde's daylight side expanding as it warms up. The low is caused by the contraction of air on the night side.

Obviously, Arde has no tornadoes or hurricanes.

game with an inflated circle (which may as well be called a "ball") and a pole set into the ground.

Two teams of three Punizlans each were competing with each other to keep the ball in play as long as possible without letting it touch the ground. At either side of the game, Punizlans stood in two collapsible triangular stands, cheering on their respective teams. Yendred could see nothing until he climbed a ladderlike affair at the back of the western stand. He would have to wait until the game was over before continuing his trip, but explained to us that he had the legal right, as a traveler, to have the game halted and the pole removed so that he could walk over all the spectators and players on his way to the city. But this would certainly cause a great deal of ill-feeling, and so he decided to wait.

The three of us watched the game on the display screen, appreciating it (so we thought) in a way that no Ardean could. Rarely had an athletic competition here on Earth so captured my attention. The ball would go zooming over the pole and a member of the team on that side would try to tip it up into the air. Usually this happened. In fact, there was no way a player could miss the ball if it was within reach; it was really a question of deflecting it properly toward a teammate. Several times we saw the play called when a player struck the ball twice. Apparently each team could keep the ball in play on its own side as long as it wanted to, but, sooner rather than later, it would send the ball speeding back over the pole, hoping to surprise the other side. Since the pole was opaque, this gave the game a nice air of unpredictability—at least for the players.

"Hey," remarked Chan. "That's just volleyball."

"Invisible volleyball," chimed in Alice under her breath. "They can't see a thing!"

We dubbed the game "Punizlan volleyball."

When the game was over, the stands were collapsed and the pole taken down, and Yendred got to walk over all the spectators in the western stands as well as the three members of the western team. These were returning to their houses in the west. Yendred followed the players and fans from Is Felblt toward the city.

Chan, currently at the terminal, seemed quite excited about Is Felblt and what it might contain.

▪ HOW MANY NSANA LIVE IN IS FELBLT?

ABOUT 6,000 NSANA LIVE THERE. DO YOU LIVE IN A CITY?

▪ YES. IN OUR CITY THERE ARE 275,000 HUMANS.

THAT STAGGERS MY IMAGINATION. IS IT THE CAPITAL OF YOUR WORLD? WHAT IS IT CALLED?

▪ IT IS NOT THE CAPITAL OF ANYTHING. IT IS CALLED ''LONDON.''

DO YOU THEN HAVE EVEN LARGER CITIES?

▪ THERE ARE MANY CITIES ON EARTH LARGER THAN THIS. ONE OF THESE IS ALSO CALLED ''LONDON'' AND IT HAS MORE THAN 7,000,000 HUMANS.

I AM THINKING THAT YOU HAVE SO MANY HUMANS THAT YOU CANNOT MOVE. ON ALL OF ARDE WE HAVE COUNTED ONLY 37,739 NSANA.

▪ WE HAVE MORE ROOM THAN YOU DO. WHY ARE YOU SLOWING DOWN?

I AM ALMOST IN IS FELBLT BUT I HAVE A PAIN.

▪ WHAT IS WRONG?

MY BRAIN HAS AN ACHE. PERHAPS WE SHOULD NOT HAVE TALKED ABOUT SO MANY HUMANS.

▪ I AM SORRY ABOUT THAT. I WAS GOING TO TELL YOU ABOUT A PLACE CALLED ''CHINA'' BUT PERHAPS WE WILL TALK ABOUT IT LATER.

YES. LATER.

We agreed to end the contact there and to restart in two days, giving Yendred time to rest at his relative's house in the city.

After ending the contact, it was midnight once again but we sat for a while discussing possible Ardean sports. Footraces would be pointless unless one were permitted to topple opponents, bowling would be a complete bore, and football a fiasco. Perhaps tennis might work.

5

City Below Ground

Thursday, June 5, 2:00 P.M.

It was early afternoon in Is Felblt when, two Arde-days later, we caught up with Yendred in the middle of the city. Of course, one should say "above the middle of the city" since all of its buildings are below the ground. An Ardean looking down on the city from the air would see nothing to indicate a metropolis except large numbers of fellow citizens going from one entrance to another—rather like looking down on a one-dimensional ant colony, perhaps.

We had been keenly anticipating Yendred's visit to the city and I am certain that each of us had formed a private mental image of what to expect. I personally had fancied Ardeans climbing all over each other in a chaotic scramble. It just never occurred to me that the Punizlans would treat themselves as we treat our cars, organizing all their comings and goings on the city's surface into a single, regulated "main street." The mechanisms making this possible were stepped depressions in the surface which we came to call "traffic pits." The pits were evenly spaced along the surface of the city, usually every thirty meters or so, with several buildings between.

A number of Punizlans would enter one of these pits from the same direction, unhooking a rope which covered the pit and hooking it back up again when all were safely inside. Presently, a number of Punizlans would come from the other direction and walk along the rope over the pit. In doing so, they seemed like casual circus aerialists and we had to remind ourselves that it was impossible for them to slip off the rope. When the rope had been crossed, those in the pit would unhook it and continue on their way.

Following Yendred, who was heading west at the time, we saw him enter every second pit and soon deduced that these pits were for the exclusive use of west-moving pedestrians. The remaining pits were for those moving east. Only one problem remained.

▪ HOW DO YOU KNOW WHEN TO LEAVE YOUR PIT?

IT IS A CROSSOVER, NOT A PIT. A HIGH GONG SOUNDS FOR WEST TRAFFIC AND A LOW GONG SOUNDS FOR EAST TRAFFIC. WHEN WE HEAR THE HIGH GONG, WE OPEN OUR CROSSOVER, PASS THE NEXT ONE, AND ENTER THE ONE AFTER THAT. WHEN THE LOW GONG SOUNDS WE CLOSE THE CROSSOVER.

Yendred encountered one Punizlan who was not bound by these rules. Here again we found ourselves suddenly dumbfounded. In the course of Yendred's walk from one traffic pit to the next, a strange figure floated onto the screen—a Punizlan hanging on to two ropes by handles, the ropes going straight up off the top of our screen.

His feet touched the ground just in front of Yendred's group and he gave a little bound which carried him gracefully over their heads to land on the other side. Scanning upward, we found the cause of his lightness.

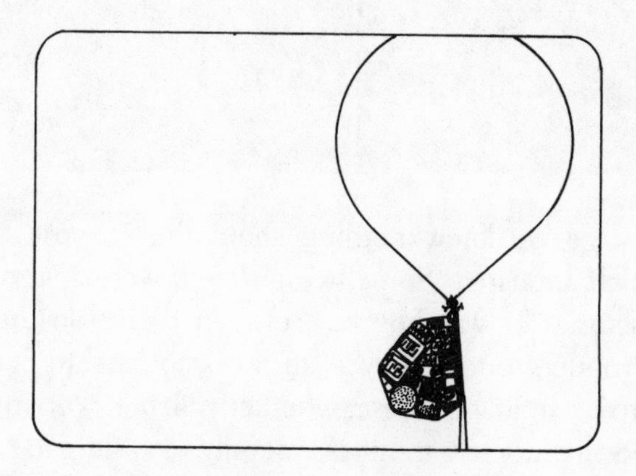

Here was the Punizlan equivalent of a delivery van. Suspended from a balloon was a parcel of goods and the Punizlan himself. The balloon, which, we supposed, contained some lighter-than-air gas, apparently gave the Punizlan and his delivery bag enough buoyancy to overcome almost all their combined weight. It looked rather enjoyable.

■ WOULD YOU LIKE TO OPERATE A DELIVERY BALLOON?

NO. IT IS HARDER THAN IT SEEMS AND VERY TIRING TO THE ARMS. BESIDES IT TAKES CAREFUL JUDGMENT TO KNOW HOW MUCH GAS TO LET OUT AFTER EACH DELIVERY. IT IS A SPECIAL TRADE.

Although we could never scan very far from Yendred without endangering the Earth/Arde link, we chanced it shortly after this episode in order to get a look at the city of Is Felblt as a whole. To achieve this, we also changed the scale of our display until Yendred was little more than a dot on our screen. The view was a bit like looking at an Earth city upside down.

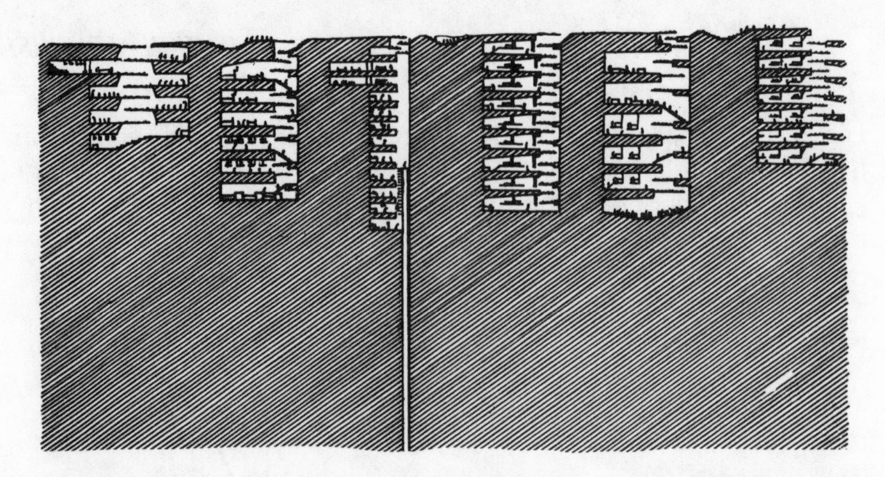

At this stage we knew nothing about the purpose of the various structures which appeared. Some, we guessed, were apartment buildings and others factories or warehouses. Between them (on the surface) and within them (underground) flowed an incessant stream of tiny Ardeans. We could zero in on any of these, whether walking, working, or sleeping. In fact, we did spend a few minutes carefully recording a mating scene in which a couple "enjoyed" an egg together. Watching all this activity, whether close up or at a distance, one felt like the ultimate voyeur able to view everything with a completeness impossible, and even incomprehensible, to the Ardeans themselves. Who might be thus watching us three-dimensional beings as we rush about our equally insubstantial three-dimensional space?

Yendred stopped before one of the entranceways. Those following him waited, flapping their arms with seeming impatience while Yendred slowly descended the stairway to what he described as his uncle's house. He had been staying with his uncle these last two days and would depart on the morrow.

- IS THAT YOUR FATHER'S BROTHER?

NO. IT IS MY MOTHER'S BROTHER. SHE GAVE ME A LETTER FOR HIM. I HAD NEVER MET HIM BEFORE. HE IS A PRINTER BUT I DO NOT LIKE HIM.

- WHY DON'T YOU LIKE HIM?

HE IS ALWAYS IN A HURRY AND NEVER LISTENS. HE THINKS HE KNOWS EVERYTHING.

His uncle's house was quite impressive, having five floors. On the top floor was a kind of shop, which Yendred now entered via the swing stair. There were no less than three lights—all with separate batteries—a lot of shelves stacked with strips of some material which could only be described as paper and, at one end of the room, a machine which looked like a primitive printing press.

THE PRINTING PRESS

The printing press is essentially just a three-part table consisting of an ink tray on the right, a press bed in the middle, and a roller stand on the left. A weight is rolled over the press bed and back to the roller stand, a cover swings from the ink tray over the press bed and back again, and the press bed contains a single line of type.

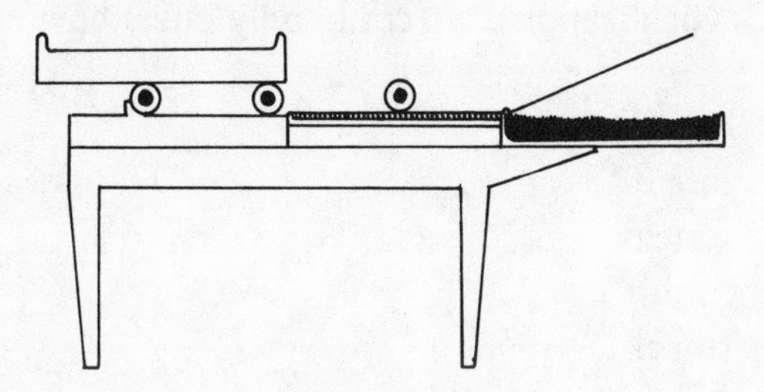

The printing procedure consists of six steps, repeated over and over, once for each strip of "paper" to be printed.

(1) Stroke the clean upper surface of the ink-tray cover with a clean stick, pressing the bottom of the cover into the ink-saturated fibers.
(2) Lay a strip of blank paper over the type (which is already inked) on the press bed.
(3) Swing the cover over onto the paper.
(4) Roll the weight over the cover, printing the paper and simultaneously inking the roller. Roll the weight back again.
(5) Swing the cover back over the ink tray and remove the printed page.
(6) Roll the weight across the press bed once again, inking the type. Then roll it back.

Essentially it is just a table with a weight, rollers, roller stand, press bed, and ink tray. Yendred had watched his uncle printing on the previous day and now described the operation in some detail, going through various motions as he did so.

All of this was quite interesting, in a way, and illustrated a strong Punizlan characteristic which was certainly shared to a degree by Yendred—namely, a fascination with mechanical devices and detailed procedures. But we were much more interested in seeing the type. This might well be our only chance to record the Punizlan alphabet.

Yendred held up each letter of the alphabet in turn, the raised type making it immediately clear that each character was composed of dots and dashes, not unlike our Morse code. Yendred pronounced each letter as he held it aloft. Ardeans speak by swallowing air into their stomachs and then forcing it out through their jaw muscles in little bursts. In the rush to note the shape of his vocalizations, we could only guess how each of the letters might sound.

▪	hiss
▬	squeak
▪ ▪	"shhh"
▪ ▬	smack
▬ ▪	"tsk"
▬ ▬	"pssst"
▪ ▪ ▪	whoosh
▪ ▪ ▬	bubbling noise
▪ ▬ ▪	burbling noise
▪ ▬ ▬	"pop"
▬ ▪ ▬	gurgling noise
▬ ▪ ▪	gargling noise
▬ ▬ ▪	"phtt"
▬ ▬ ▬	choking sound

We wanted next to ask Yendred about Ardean books and literature generally, but his uncle was now coming up the stairs to the shop.

▪ EAVESDROP

YOU A PRINTER TO BE WOULD LIKE?

IT A WONDERFUL PROFESSION IS. YOU ALL WHICH READ IS MAKE. YOU WITH TEACHERS A PARTNER ARE.

YOU TRULY SPEAK. IF PRINTERS NOT WERE THEN CIVILIZATION COLLAPSE WOULD. I FROM THIS BESIDES MUCH MONEY GET. BUT DOWN COME. WE EAT.

The two descended the stairs to the kitchen, where a family of four others was waiting.

The meal was served, bowls and plates being passed overhead from the stove along the row of family diners. We noted once again how slowly Ardeans ate, chewing their food carefully and passing it back into their stomachs. When their stomachs were full, they would slump contentedly on their stools for a long time and then cock their heads to one side, spitting up all the solid particles into a container kept near their feet. The sight of this practice never failed to revolt Alice.

Following one of these lulls, Yendred's uncle questioned him about his journey:

YOU THINGS IN VANIZLA THAT HERE ARE NOT ARE SAY?

YES AND THAT WHY I GOING AM IS.

I WHAT NOT ASK WILL. VANIZLA A LAND OF BACKWARD IGNORANT NSANA IS. MACHINES THEY NOT HAVE. ELECTRICITY THEY NOT HAVE. ROCKETS THEY NOT HAVE. BOOKS PRINTED NOT ARE BUT BY HAND WRITTEN. VANIZLANS IN ONE PLACE TO SIT LOVE AND NOTHING DO. WE EVERYTHING HAVE. VANIZLA NOTHING HAS.

BUT THEIR PHILOSOPHY DEEP IS IT NOT IS?

THIS A PHILOSOPHY OF FOOLS IS AND A RELIGION FOR THE IGNORANT.

BUT OF THE BEYOND WHAT?

OF THE BEYOND NOTHING. A FICTION THE GULLIBLE TO ENSNARE. NOTHING THERE IS WHICH NOT HERE IS. THEIR JEALOUSY OF OUR PROGRESS THEM CRAZY HAS MADE. THEY A SECRET TO HAVE PRETEND AND THIS EVEN TO THEMSELVES.

From this point on, Yendred made only vague, noncommittal replies to his uncle's queries and comments. Fortunately, the conversation turned shortly after this to a discussion of a gathering which his uncle was hosting that very evening, a sort of party in Yendred's honor. Presently, the family removed the dishes, collapsed the stools back into the floor, and retired to the rooms downstairs—except for the uncle, who went upstairs to do some printing. We watched the uncle work, deftly managing sheets, cover, and rollers, and we pumped Yendred for more information about everything from book construction to Ardean literature.

When the individual pages of a book are printed, they are placed inside a cover and glued to the inside of the spine. The resulting object looks exactly like the cross-section of an Earth book.

Although Punizlans and Vanizlans employ the same alphabet (and, for that matter, speak the same language), Punizlans read from left to right while Vanizlans read from right to left. In fact, it is not quite accurate to say that the two nations use *exactly* the same alphabet, for the characters in one are mirror images of the characters in the other.

On a number of occasions we saw Punizlans reading books or documents by laying them open upon a lower hand and turning pages with the upper hand. If the Punizlan held the book thus on the east side of his body, he would scan the pages from east to west, steadily declining his head downward as he read. On the other side of his body he would still read from east to west but inclining his head slowly upward as he did.

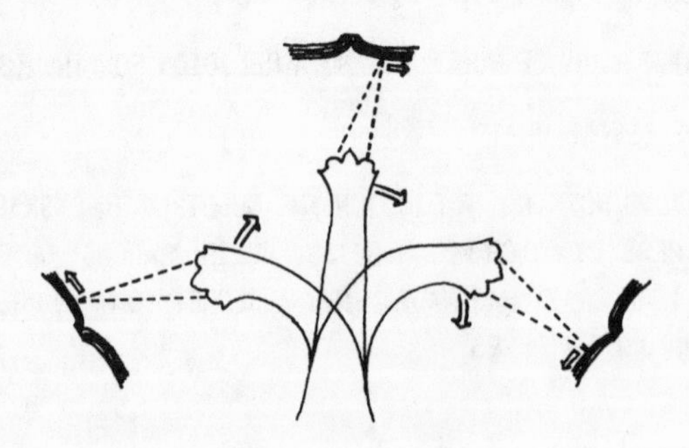

Vanizlans, of course, read everything in the opposite direction. Ardean literature, both of Punizla and Vanizla, has the pithy, epigrammatic quality of poetry. This is almost surely due to the fact that each page of an Ardean book can hold about one sentence.

The weightiest tomes, by Ardean standards, would amount to little more than pamphlets on Earth, at least in terms of their total number of words. Yet, we must believe, the weight of their meaning is no less. In any event, Ardeans take extraordinary care with everything they write, and try to say much with few words. Another consequence of restricted dimensionality on Ardean literary life is that there are far fewer titles in print at any time than is the case on Earth. Moreover, because of Arde's much smaller population, far fewer copies of each title are printed. No home on Arde contains much over a dozen books and even libraries generally contain less than a thousand volumes.

Punizla and Vanizla both suffer from the same shortage of space for books, and each has solved this problem in a different way. In Punizla, only the great works of science, mathematics, and philosophy are permanently enshrined in the few great libraries. Since Punizlans also enjoy light reading, books of this nature must also be accommodated. Thus, works of adventure and romance are regularly printed in Punizla but, just as regularly, they are rounded up to recycle the paper. This must cause occasional feelings of bitterness in the minds of their authors, but there seems no other way of encouraging such literature without Punizla becoming inundated with books.

The Vanizlans, for their part, only rarely produce anything new, it being considered presumptuous to attempt improving on a work of metaphysics by an old master. Any new work must pass a five-year trial period in single-copy form before it is approved. Most often, such trial works are returned to the author after five years, having failed to pass stringent tests of relevance, depth, or mystical efficacy.

Shortly after Yendred's uncle completed his printing, fellow Ardeans began to arrive at his home and make their way down the entrance stairs one by one. The entire bottom floor had been made over for the party, with bowls of food and bottles set out on a table at one end. Beside the table stood Yendred's aunt, who would pass these refreshments out to the guests as the party progressed.

Several of the guests brought bottles with them, bottles which were tightly capped but apparently empty.

▪ WHY ARE THE BOTTLES BROUGHT BY THE VISITORS EMPTY?

THESE ARE BOTTLES OF HRABX, THE GAS WE MUST BREATHE TO STAY ALIVE. WITH SO MANY NSANA IN THE HOUSE WE WOULD ALL FAINT UNLESS WE HAD EXTRA HRABX. AS A CUSTOM OTHER GASES ARE MIXED WITH THESE. SOME GASES HAVE A STRANGE EFFECT ON OUR MINDS. WE BECOME VERY FRIENDLY AND THINK LESS AND LESS. IT IS A GROWING HABIT IN PUNIZLA TO BRING THEM TO SUCH GATHERINGS.

When the party was in "full swing," so to speak, the guests stood between host and hostess on the bottom floor; they stood in the same order in which they arrived and would have to preserve that order for the duration. Each guest had the choice of only two (or, possibly, four) conversational partners, but, judging from the frequency of their vocalizations, had much to talk about nevertheless. One by one the bottles were opened and the gases allowed slowly to escape. Talking grew more continuous. Food and beverages were passed overhead from guest to guest.

In attempting to use the Eavesdrop facility, all we got was very strange output, perhaps the effect of everyone talking at once:

REY $YY TTUV PBARQ# TY XDNDXN

But if we focused on a single Ardean and then used Eavesdrop, a comprehensible pattern emerged. After two hours, all the bottles had been opened and the guests were all talking, swaying back and forth, and even seeming to stumble a little. Luckily it is not easy for a two-legged being to fall over in a two-dimensional world.

At this point, we noticed that Yendred was talking as much as anybody, and to no one in particular, his head pointing straight up. The two guests beside him were silent, apparently listening. Then the guests beside these became quiet. Before long, Yendred was the only one talking. Imagine our horror when we decided to eavesdrop on Yendred and watched the following lines emerge from the printer.

DELIGHTFUL IS. YOU WHAT DO KNOW? YOU NOTHING KNOW. I FRIENDS IN PLACES OTHER HAVE. I INVISIBLE FRIENDS FROM EARTH HAVE. THEY WITHIN THE CIRCLE OF CIRCLES DWELL AND ME THE GREATEST SECRETS HAVE TOLD. YOU OF THESE THINGS NOTHING KNOW. I YOU NOW ABOUT THE THREE DIMENSIONS WILL TELL.

▪ YNDRD YOU SHOULD NOT SAY THOSE THINGS.

AH! THEY THERE NOW ARE. YNDRD YOU SHOULD NOT SAY THOSE THINGS. MY EARTH FRIENDS ME THE SECRETS NOT TO TELL ASK. THEY YOUR BLOOD AS YOUR FACE EASILY CAN WATCH. THEY EVERYWHERE ARE.

The uncle, perhaps sensing some strange madness compounded with his nephew's intoxication, had Yendred brought to the foot of the swing stair; guests crowded ahead of him to the other side of the platform.

One of the guests helped him up the stairs to a bedroom, where Yendred fell over onto a bed, his arms draped carelessly along his side. The guest raised an arm so that it lay over his head and returned down the stairs to the party. Yendred lay on the bed, inert. Before we could even begin to terminate the 2DWORLD program, the screen began to wobble and the printer delivered a parting message:

EARTH SPIRITS. GUIDR MT TJI PREDT/BU . . .

We shut down the 2DWORLD program and returned to our normal existences to sleep, to eat, and to live for a time in a world which, for me at least, seemed a few degrees less real than ever before.

Friday, June 6, 2:00 P.M.

It was the middle of the morning in Is Felblt when once again we communicated with Yendred. He had left his uncle's after a hurried breakfast and a not entirely fond farewell. Had anyone taken seriously the things he had said about his "friends from Earth"?

WHAT THINGS? WHAT DID I SAY AND HOW DID YOU HEAR ME?

It was time to reveal our ability to eavesdrop. I much preferred the risk of inhibiting Yendred's conversations to having our existence revealed. We told him how we could listen in and quoted liberally from the previous night's transcript. Yendred was silent for a long time.

> PLEASE FORGIVE ME. I DID NOT KNOW WHAT I WAS SAYING. THAT DREADFUL GAS.

He went on like this at some length, asking us not to desert him on account of his heedlessness. By degrees he grew more reflective and ended by remarking that this explained why his uncle had stated over breakfast (violating mealtime etiquette in so doing) that the "circle of circles" was no fit concept for a serious Punizlan; Yendred was welcome to travel to his doom among the ignorant Vanizlans.

As if struck by a sudden worry, Yendred reached into his pocket, took out his money, and began to count it. Punizlan currency is based on the jeb, a small "coin" made of jebb and shaped like a narrow rectangular chip. There are one-, two-, four-, and eight-jeb coins, as well as some flexible strips in denominations of 64 and 128 jeb. He had only enough for a balloon ride covering part of the distance to Vanizla. How could he possibly walk the rest of the way?

> ▪ WHERE IS THE NEXT CITY?
>
> THERE ARE SOME SMALL CITIES AHEAD BUT THE NEXT LARGE CITY IS SHEMS TRABLT. IT IS ABOUT 100 FSADS FROM HERE.
>
> [100 fsads = 420 kilometers]
>
> WILL YOU WALK ALL THE WAY?
>
> ▪ NO. I WILL USE A TRAVEL BALLOON. I HOPE TO BUY ONE TODAY.

This turned out to be a balloon similar to that used for carrying heavy objects or for making deliveries. Suddenly the prospect of such a vast distance to be covered appeared less daunting.

The city below Yendred's feet showed few signs of petering out, and we used this opportunity to explore some of the structures which Yendred crossed. These included a school, a factory, the Punizlan parliament building, and something like a steel mill, which I will presently describe.

First, however, it is necessary to explain a few hard facts of Ardean architecture. Most underground structures are organized vertically because it can be dangerous to tunnel horizontally any great distance.

A tunnel invariably creates a large overhang of unsupported soil or rock that is certain to collapse if the tunnel is extended too far. In Punizlan history, cave-ins were the leading cause of accidental death, at least until the door/wall was invented. This device made it possible to build tunnels of much greater length in spite of certain drawbacks: although it will support tremendous weights, it requires a good deal of energy to open or close and so is useful only when access to the space beyond is not frequent. Thus one often finds door/walls used with horizontally extended storage rooms.

THE DOOR/WALL

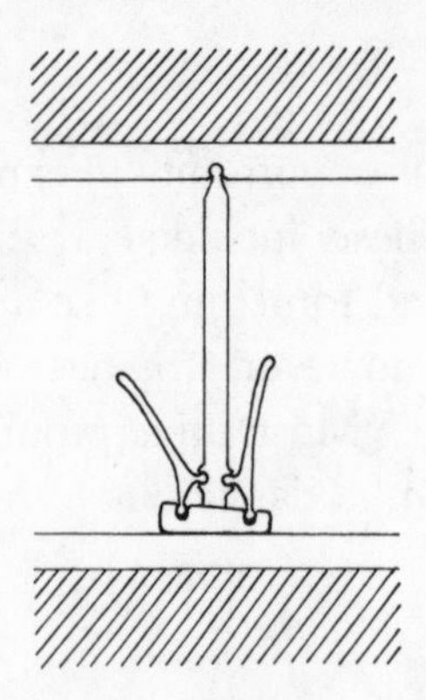

This strange but useful device has just four moving parts: a hinged column, two levers, and a shoe. The shoe fits into the bottom of the column at an angle so that it acts like a wedge. It is pushed under the column in order to support the ceiling and it is pulled away to allow the column to hang freely.

Approaching the door/wall from the left, an Ardean pushes the lever, pulling the shoe out from under the column. It is then possible to swing the column to the right as the Ardean walks under it. In order to return the door/wall once more to its supporting role, the Ardean now pushes against the lever on the other side of the door/wall.

An Ardean passing under the door/wall in the opposite direction does precisely the same things, but pulls the lever instead of pushing it.

Door/walls, moreover, must be used in pairs so that one is bearing the weight while the other is being opened or closed. We found just such a pair in a Punizlan school.

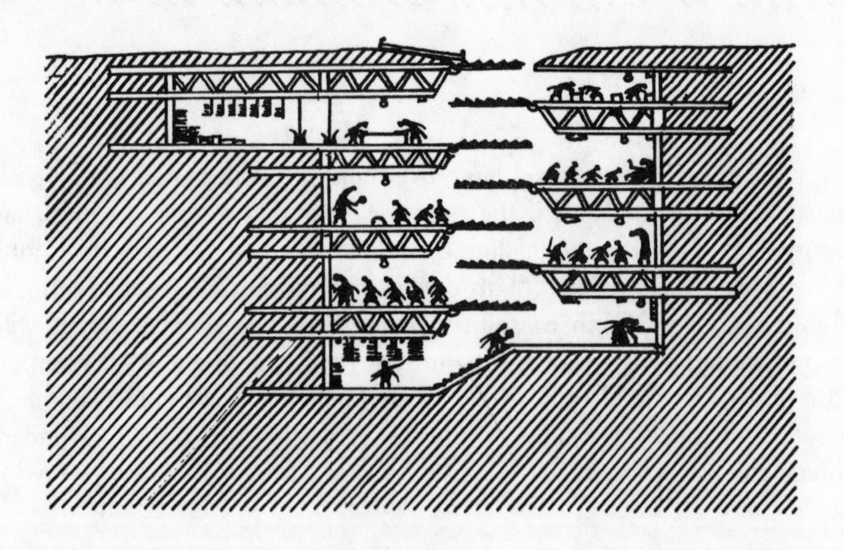

Yendred was about to cross the entrance to the school when the students began to arrive from both directions at once. Rather than allow them all to walk over him (which he would have to do according to Punizlan crossing rules), Yendred elected to enter the school in the hope that there would be a nook of some kind where he could dodge the youthful traffic. This he found in the staff work area just below the entrance.

The students and, indeed, the teachers had all timed their departures from their homes that morning so as to arrive in the correct order. The principal or headmaster was already hard at work in his basement office. The school library was also on this floor.

Each class entered the school in turn, preceded by its teacher, the first class going to the second-to-the-bottom floor by a succession of swing stairs, the next class going to the floor above that, and so on. Before the staff arrived, Yendred made his way to the surface and continued his journey. We lingered on the scene for a few minutes, watching with fascination as one of the teachers began a lesson on a sort of blackboard at the front of his classroom. He started writing at the top of the board and, as he wrote, crouched lower and lower in order not to block his students' view.

Yendred next passed over a glue factory, which, he said, gave off a very bad smell. We briefly examined several floors containing vats and bins of chemicals before catching up with Yendred as he passed over some "apartment buildings." The next structure of interest was a sort of factory containing a very deep shaft in which a movable column slid up and down. Small workrooms gave off the shaft from the surface all the way down to a depth of 150 meters.

▪ WHAT BUILDING IS THIS?

THEY MAKE BATTERIES HERE.

▪ HOW DO YOU KNOW?

THERE IS A SIGN ON THE ENTRANCE STAIR WRITTEN ON THE STEPS. CAN YOU NOT SEE IT?

Attached to the column was a series of ledges on which sat a number

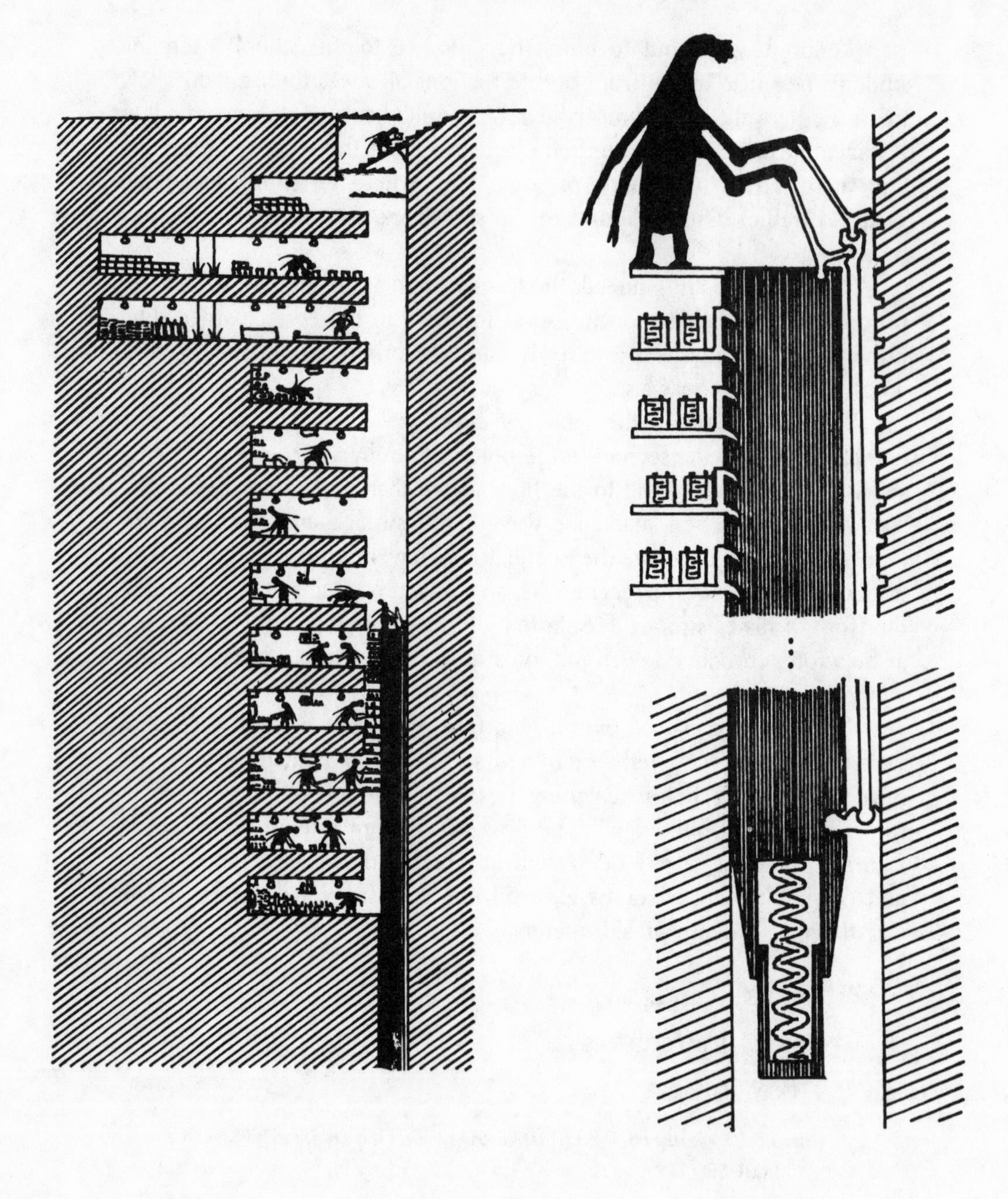

of half-completed batteries. The workers in the rooms beside the ledges would remove the batteries, add some component or other, and replace them on the ledges. An operator standing astride the column would then pull two levers and the column would rise until the ledges were brought up to the next set of workrooms.

The mechanism attached to the column, enabling the operator to raise or lower it, consisted of a long rod extending to the base of the column where it ended in a sort of latch which could be raised or lowered to allow steam from below to escape. As steam escaped, the column would sink slowly down the shaft. We began to scan down the shaft to discover the source of the steam, when the picture began to fade and waver on the screen. Chan, designated recorder for this session, quickly snapped another Polaroid photograph.

We got back to Yendred as quickly as possible in order to strengthen the Arde/Earth link before it dissolved entirely. He had heard of this factory and claimed that the steam was generated in a vast underground reservoir where heat from the planet's interior acted upon water poured down the shaft at intervals.

Yendred came next to a rather impressive structure which turned out to be the Punizlan parliament or house of assembly. The building was almost empty of Punizlans at this time, but it did not take us long to find the central hall of debate (based on Yendred's description) nor to imagine how it would look when the house was in session.

There are eight regions in Punizla and each one elects a representative from either of the two official parties. The eight representatives sit in this small assembly presided over by the leader of the party that obtained the largest popular vote in the election.

The Punizlan names for the two parties meant nothing to us so, based on their respective platforms, it seemed fair to call one party the Planners and the other the Naturals.

The Planners wish to define Punizlan society completely through laws which govern social and economic behavior, erecting a massive bureaucracy to oversee their everyday application. Their vision of Utopia involves a society where almost nothing is left to chance and in which all Ardeans are viewed as parts of a vast social machine. At election time the appeal of the Planners is based largely on their promise to protect Punizlans from economic, social, and medical misfortunes. They have been in power twice since Punizla became a democracy.

The Naturals believe that it is better to allow natural forces, whether social or economic, to govern Punizlan affairs, whether public or private. While conceding the need for some laws, the Naturals wish to keep these to a minimum, arguing that an excess in one direction is always corrected, sooner or later, by an opposing natural force. This balance of "natural" forces is supposed to guarantee stability in the long run, even if some unpleasant oscillations occasionally result.

On the whole, Punizlan society can be described as dynamic and progressive, although neither party has been able to deal with a slow and steady rise in the crime rate. On this subject, Yendred had some particularly interesting information: murder, theft, and rape are all on the increase.

Murder is normally committed either with poison or a knife. Even if Punizlans had guns, it would be pointless to try shooting anyone since bullets would presumably bounce off the Ardean exoskeleton. A knife inserted into the portal muscle below an armpit is usually fatal, the victim bleeding to death.

Theft of large items such as radios, hrabx bottles, and so on, is quite rare in Punizla, probably due to the difficulty of concealing such things after leaving the victim's premises. However, money and gems are favored targets. A Punizlan thief might even follow an older citizen into a traffic pit and there, even as traffic in the other direction patters overhead, force his victim down, taking everything of value from bag or pocket.

Rape is really a form of theft, in which a male Ardean, under pretense of eternal fidelity, persuades a female to release her egg. He then makes off with it to a place where he and perhaps a comrade or two may enjoy it at leisure.

Both the Planners and the Naturals are somewhat split on the question of how to deal with crime. The general drift in recent years has been oriented toward rehabilitation of the criminal, rather than punishment. Increasingly, in a desperate attempt to modify such antisocial behavior, Punizlans wish to understand the criminal mind. It seems very bad luck indeed that crime should increase the better criminals are understood.

Besides the rather serious crimes enumerated above, Punizlan society has become generally less law abiding in smaller matters as well. Young Ardeans frequently ignore passing rules and topple oldsters in order to walk over them. Fewer Punizlans obey the confinement law in failing to check against the possibility of accidentally confining a fellow citizen in a dead-air space. Such spaces are created very easily, as, for example, when one leaves a plank lying across a stairway; below, the unwitting victims become faint and pass out.

In contrast to these negative elements, we observed that it must be fairly easy to search for criminals on Arde. In fact, it would not be totally out of the question to conduct a house-to-house search of the entire planet.

After this long discussion about Punizlan government and society, Yendred was rather anxious to continue his trip and we reluctantly left the center of Punizlan power, making notes about possible questions to ask Yendred at some future time.

The afternoon was now well advanced, and Yendred walked with some haste, apparently anxious to be out of the city by nightfall. In the laboratory it was nearly midnight, and Edwards, who had been trying to learn more about Vanizla, took over from Ffennell at the terminal. These two formed an interesting pair of students, typifying two complementary mentalities. Both had transferred into computer science from mathematics and still bore the stamp of that discipline, although in different ways. Ffennell's mind was quick, like a passenger train that sped over some mental landscape. But Edwards' thoughts gathered momentum slowly, like a freight train pulling out of some complicated marshaling yard. Where Ffennell might quickly find a proof for a theorem presented to him, Edwards would someday find the theorem itself.

Yendred now told Edwards about Vanizlan society, contrasting its form of government and social structure with that of Punizla. Vanizla was ruled by a king and eight viceroys. Apparently undemocratic, the country was at least stable, albeit "backward and ignorant" as Yendred's uncle might have said. Vanizlans value their system more than they do technological and social progress, the result being a stagnant society rarely, if ever, visited by change. Edwards asked whether Yendred could be more specific about what he sought in such a place.

▪ CAN YOU TELL US WHAT YOU EXPECT TO FIND IN VANIZLA?

THAT IS VERY HARD TO DESCRIBE. I HAVE READ IN CERTAIN BOOKS ABOUT A KNOWLEDGE WHICH THE VANIZLANS POSSESS. AT LEAST SOME OF THEM POSSESS IT.

▪ WHAT SORT OF KNOWLEDGE?

VERY OLD KNOWLEDGE CONCERNING WHAT LIES BEYOND.

▪ WE LIE BEYOND, DON'T WE?

YOU ARE NOT HERE SO YOU MUST BE BEYOND. BUT NOTHING YOU HAVE SAID SO FAR SOUNDS LIKE MY BOOKS.

▪ WHAT DO YOUR BOOKS SAY?

THEY SPEAK OF A PRESENCE WHICH INHABITS THE BEYOND. IT IS CALLED SIMPLY ''THE PRESENCE'' OR SOMETIMES ''THE SOURCE.''

Edwards hunched toward the screen, glancing every few seconds between it and the printer. Ffennell, preoccupied with a question about the Ardean method of counting, nudged Edwards.

"Don't forget to ask him to count."

"Not now." Edwards turned back to the screen.

▪ WHAT IS ''THE PRESENCE''?

I WAS HOPING THAT YOU WOULD TELL ME THAT.

▪ WHAT VANIZLANS POSSESS THE KNOWLEDGE OF BEYOND?

THERE IS ONE CALLED DRABK I HAVE BEEN HOPING TO MEET.

Ffennell took up a seat outside the circle of students. He glanced at the screen from time to time as he made notes on a pad. The conversation with Yendred continued for only a few more lines when suddenly a large factory of some kind edged its way onto the screen. This was a large open pit scarring the surface of Arde. Yendred would have to wait until the factory suspended operations for the day. It was the last large structure in eastern Is Felblt.

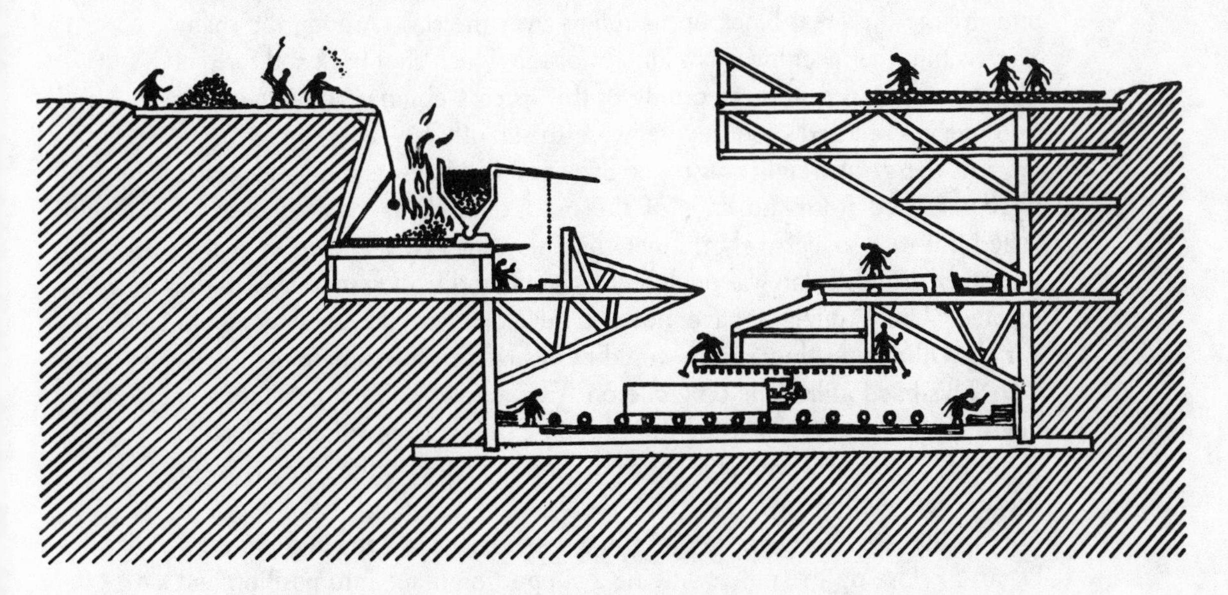

The factory was a "hadd mill," hadd being a greenish powder found in the sediments washing down from Dahl Radam. When heated to very high temperatures, at least by Ardean standards, it gives off a smoky vapor and turns into a bright, heavy, viscous liquid which can be squeezed or beaten into various useful shapes as it cools and solidifies. This sounds so much like our iron that I am tempted to think that there are elements in the Planiverse which parallel ours very closely.

A dozen Punizlans worked at the mill, an operation which consisted of a blast furnace on the upper level and a rolling mill below. On the east side of the mill, a bellows operator drew a long pole up and down by means of a rope. This fanned a large fire below a vat of hadd powder. A stoker periodically threw a shovelful of fuel over his head and onto the bellows rope, whence it bounced down into the firepit. When the hadd was liquefied and floated to the surface of the crucible, an operator pulled on a rope which tilted the crucible, pouring liquid hadd down a succession of ramps into the mill below. Here the slab cooled for a while before workers in protective clothing activated a most amazing machine. Attached to the side of an immense block sitting on rollers, the machine began to move inside and a rod, projecting from the machine to a rack above the block, began to wag back and forth, engaging teeth in the rack and driving the great block on its rollers over the slab. Among the many parts within the machine was a kind of piston which shot back and forth at unbelievable speed. Our recording of this "steam engine" was imperfect, but later in Yendred's journey we were to see others.

As the great weight was rolled back and forth across the hadd, plates were removed from the ends of the expanding slab, enabling it to be rolled out into successively thinner shapes. After it was cooled by water, the now hardened slab was enclosed in a rope and hauled manually to the surface. The Punizlans have not yet developed a successful crane or winch. Otherwise, they surely would be using it in this situation.

This hadd mill is the only one on Arde, easily producing a sufficient quantity of slabs and bars for the 20,000 Punizlans and their industrial economy. They are even able to export surplus hadd to the Vanizlans for use in their modest underground buildings.

In the late afternoon, one of the workers struck a gong and the crew began to close up the mill, swinging a supporting brace into position and sliding a long trusswork cover over the mill, restoring once again the smooth continuity of the Ardean surface. The workers then all lay prone to let Yendred walk over them. He would encounter a depot in ten minutes.

We had decided earlier, at Ffennell's urging, to share some take-out food for supper in the laboratory. Ffennell, naturally, would have to pick it up with his car, and Edwards volunteered to go with him. The two left the room arguing about binary numbers.

"Look, I'll just ask him to count on his fingers. He only *has* two on each hand. Then we'll know."

It frequently happened during our contact sessions that I would go without a cigarette for hours on end. The urge for one suddenly came upon me at this point and I went to my office, leaving Lambert in charge.

When I got back to the laboratory, I found Lambert giggling as he typed something to Yendred. Alice looked very upset.

▪ YNDRD. WE ADVISE YOU TO TAKE HER EGG.
WE WANT TO SEE HOW YOU DO IT.

"What's going on?"

Lambert froze at the terminal.

Alice quickly brought me up to date. Yendred had crossed the hadd mill and several houses, coming at last to a lone house with a depot beside it. As Yendred passed the swing stair of this house, a female had addressed him from below the stair.

▪ EAVESDROP

YOU MY EGG TO BUY DO WANT? IT A BEAUTIFUL BLUE IS AND VERY
LARGE AND GOOD TO SIT UPON.

YOUR EGG WITHOUT DOUBT BEAUTIFUL IS BUT I NO MONEY HAVE.

"He was lying," said Lambert. "He had plenty of money."

▪ YNDRD. BUY THE EGG. WOULDN'T YOU LIKE TO SIT ON IT?

YES. VERY MUCH. IT ATTRACTS ME STRONGLY. BUT IT IS NOT RIGHT
AND I MUST BE ON MY WAY.

▪ YOU HAVE PLENTY OF TIME.

Even before I came to the last sentence in this incredible exchange, I was truly angry. Lambert was already cringing.

"Do you think this is a video game? Will you get five hundred points if Yendred takes the egg? You have been interfering with Yendred's actions, perverting his natural thought patterns and bending his whole world to your own selfish ends. Perhaps it's time you left for the night."

Lambert looked rather sad and sheepish as he collected his things and left the laboratory.

▪ DEWDNEY HERE YNDRD. JUST DO WHATEVER IS MOST NATURAL FOR YOU.

Yendred said goodbye to the young female and continued on to the depot.

He entered and went down the stairs to a sort of store, where he withdrew one of the bills from his money pouch, along with two of the jeb coins, and purchased a travel balloon. This consisted of a long rope arranged into a layered package, a handle with slots to receive the end of the rope, and a small container of pressurized gas. Yendred also bought some food items and placed these in a string bag.

Thus equipped, Yendred climbed to the surface late in the Ardean afternoon and began a journey that was to last two of our weeks, a journey filled with adventure and a strange encounter. During this period, our computer was also fated to suffer a serious breakdown.

6

The Trek

As a professor of computer science, I find myself so close to computers so much of the time, so intimately involved in hardware design, programs, and algorithms, that I rarely step back from these magnificent machines and whistle in amazement or reverently mutter the scientific equivalent of "gee whiz." This is a common failing of most scientists, an inability or unwillingness to look at their subject with fresh eyes as a lay person or, better yet, as a child might.

Our experience with the 2DWORLD program, its sudden inexplicable behavior and the emergence of Arde as a reality from the depths of our display screen, jolted me out of my scientific rut and awakened the child within to the endless possibilities of existence. How refreshing it was to break free of the mold cast for me by routine minds thinking routine thoughts. Why should I care whether such minds would never be convinced that another universe, besides our own, exists?

Saturday, June 7, 11:00 P.M.

There before me Yendred traversed the surface of Arde on our screen and to my right was the computer itself, bathed in the light of the Planiverse, so to speak. It was (I may as well say it) a specially modified DEC System 10 with extended memory, two disk drives, and a large display scope driven in refresh mode by its own graphics memory. I can recall looking at the machine that afternoon and temporarily but deliberately forgetting everything I had ever learned about computers in order to enjoy to the full a quality of "otherness" that swept over me, challenging me to redefine within myself what it meant to be human. This alien machine, the computer, had called forth this alien universe, and now I too seemed alien to myself. How long could these contacts go on?

The manner in which Yendred was now traveling was quite new to us. Just when we had resigned ourselves to the vision of his endlessly oscillating, trundling figure, we found him alternately bounding and floating, suspended from his travel balloon.

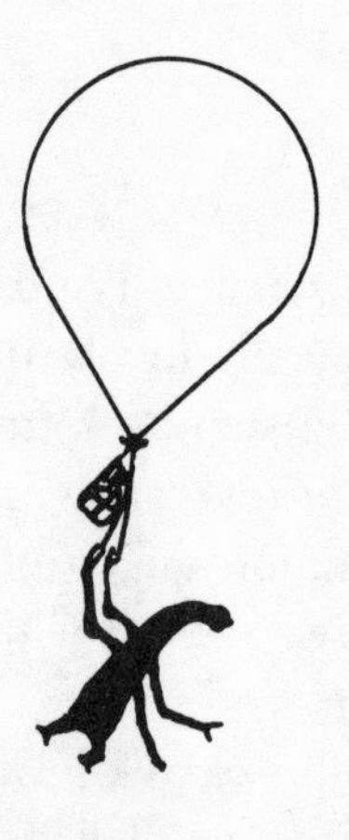

His balloon had slightly less than enough gas in it to hoist him aloft, where he would be at the mercy of the slackening but still effective late afternoon wind. It was obviously a matter of nice judgment to maximize the time between landings while insuring a low risk of being carried into the turbulent heavens. Yendred, however, seemed content with gentle leaps of five meters or so. Periodically, he would shorten the lead rope and, after landing, pull the balloon into his lee in order to stand, flex his arms, and rest them.

His answers were becoming shorter and his questions less frequent, so we decided to break contact for now. We wrote up our notes, gathered our few (rather faint) Polaroid photographs, and placed these, with the accumulated printouts, in a folder labeled with the date.

We decided it would be best to take Sunday off. Yendred would be sleeping during most of the time we had access to the computer in any event. When we assembled on Monday to resume contact, we found a "maintenance down" in progress. The systems people had found that one of the circuit boards serving main memory was behaving a bit strangely. They replaced it, but the computer would not be available again until the next day.

Tuesday, June 10, 2:00 P.M.

The ugly and artificial landscape of 2DWORLD's Astria melted into the Ardean surface as usual, and there was Yendred, not suspended from his travel balloon, but trudging along, the balloon collapsed under one arm.

▪ WE ARE BACK.

HELLO. WHO IS THERE TODAY?

▪ ALL OF US.

WHO IS AT THE TERMINAL THING?

▪ ALICE. WHY ARE YOU WALKING? YOU COULD BE SOARING ALONG WITH YOUR BALLOON.

MY ARMS HURT TOO MUCH. I HAVE BEEN WALKING ALL DAY.

As he walked, Yendred occasionally had to step over a fairly large stone. These were nearly all quite smooth and round. Alice remarked on this and Yendred explained that the stones had suffered quite a long journey, all the way from Dahl Radam, in fact. In the process, they had been worn nearly into the shape of disks. Such stones must surely roll all the way to Fiddib Har and even down the coastal slope into the submarine trench far below.

"What's that?"

Chan pointed to a peculiar assembly of rodlike objects occupying a chamber below Yendred's feet. On closer inspection, these turned out to be bones, carefully arranged into an apparently significant pattern.

▪ STOP FOR A WHILE. WE THINK WE SEE BONES IN THE GROUND UNDERNEATH YOU.

BONES?

▪ WE SEE A MOUTH AND EYES, ARM BONES, BODY BONES, AND LEG BONES.

ARE THEY MADE INTO LAYERS?

▪ YES. SORT OF.

THAT MUST BE A VERY OLD PUNIZLAN GRAVE. MANY HUNDRED YEARS AGO PUNIZLANS FOLLOWED THE VANIZLAN RELIGION AND BURIED THEIR DEAD IN THE VANIZLAN WAY.

▪ WHY ARE THE BONES ARRANGED LIKE THAT?

THEY ARE PREPARED FOR REASSEMBLY BY THE PRESENCE.

It turned out that in the Vanizlan religion it was believed that every Ardean had come from the Presence (or Source) and would someday be reabsorbed into it. Before reabsorption, all would be reconstituted from their old bones. It was therefore a matter of some importance to preserve the bones. One can readily understand the basis for this belief, since preserved bones could easily be glued to a form of the appropriate shape and a lifelike replica of the deceased reconstructed: the lifelike appearance (to an Ardean) of such a statue would naturally encourage a belief in immortality of some kind.

▪ WHAT HAPPENS WHEN THE NSANA DIE?

The resulting description by Yendred confused us at first, but gradually a rather horrifying picture emerged.

When an Ardean dies, for whatever cause, haste is made to place the body in a special bed with raised ends. Within minutes of death, the zipper muscles begin to part, one by one. Body fluid drains into the uppermost arms, causing them to balloon. The stomach everts and protrudes from the mouth. Within the tissues themselves, cell wall chambers release their hold. Cells drain, tissues break, gills disintegrate, and the stomach ruptures. Fluid leaks from the corpse and fills the deathbed, being contained only by the raised ends.

▪ HAVE YOU SEEN THIS HAPPEN?

YES I HAVE. WE WATCH AND WE SING OR WE RECITE GOOD THINGS ABOUT THE ONE WHO DIED.

▪ IS THIS IN THE PUNIZLAN RELIGION?

NO. IT IS JUST SOMETHING LEFT OVER FROM THE OLD DAYS.

After disintegration, the bones are retrieved from the deathbed and burned. The fluids are kept in a container until the next river is about to pass and then left outside on the surface.

▪ HOW LONG DO NSANA LIVE?

OFTEN TO BE 130 TO 140 YEARS OLD. DO YOU ALSO DIE?

▪ YES WE DO.

I DID NOT THINK ANYONE FROM THE BEYOND COULD DIE. WHAT HAPPENS AFTER YOU DIE?

▪ WE JUST ROT AWAY. IT IS A VERY SLOW DISINTEGRATION.

NO. I MEAN WHAT HAPPENS TO YOU?

▪ WE THINK NOTHING HAPPENS.

NOTHING?

"Be a little careful here, Alice. We don't want to damage his belief system."

▪ SOME EARTH PEOPLE SAY THAT WE GO TO HEAVEN.

Yendred then questioned us for a long time about heaven and related matters.

It was at this point that we failed Yendred. He continually seemed to expect some fragment of "knowledge of the beyond" from us. This made our position veiy delicate: to say too little might disappoint him and even destroy his beliefs, and to say too much might distort them. For this reason, I advised Alice to type the following sentence.

▪ THERE ARE MANY THINGS WHICH WE DO NOT KNOW. WE ARE NOT SO DIFFERENT FROM YOU.

I was tempted, on reflection, to add "Yendred, you are the 'beyond' for us," but didn't.

Yendred had been walking steadily east, his balloon and other provisions tucked under an arm, when he came upon a large gap in the ground. This was the entrance to a storage shed or barn in which an old chap was busily repairing a ladder.

The walls of the barn were lined with shelves on which were stored Jirri Basla heads, recently picked by the farmer's wife. She presently came into view with a sackful of the vegetables.

While Yendred waited with typical Ardean patience, the farmer's wife took her miniature harvest down the steps into the barn. Yendred knew enough about Punizlan agriculture to inform us that the farm family did not grow their own crops but waited for the crops to come to them, borne on any river which happened to come to rest in the area, in which they were licensed to harvest. This area extended several kilometers on either side of the house, which lay just east of the barn.

Ffennell, with his usual sharpness, noted that the act of harvesting was essentially a one-dimensional operation. A single sweep over the "field" was generally sufficient to gather any vegetable which chanced to be growing there—like a market garden with a single row of randomly distributed plants.

Besides harvesting the bulbs of the Jirri Basla and other vegetables that came their way, the farmers were also responsible for gathering any of the seeds which subsequently developed. They would take both the harvest and the seeds to the nearest depot, receiving money in return. The crop would then be taken to the market and the seeds handed over to a government official for later distribution on the heights of Dahl Radam. In this way, Punizlan agriculture leaves as little to chance as possible. The Vanizlans, however, follow the traditional method of allowing the seeds to make their own way to Dahl Radam.

The farmer and his wife invited Yendred to their house to share their noon meal. First, however, they had to close up the barn. The students were very puzzled, when the farmer had ascended to the surface, to see him struggling with the barn door while Yendred raised not a finger to help him. Was the farmer too old to lift the door? We had seen younger Ardeans handling larger objects. Was the door made of some extraordinarily heavy material? Why was Yendred behaving in such a callous manner?

Alice noticed that the farmer, having raised one end of the door a small distance, would do a sort of little jig with his feet and then raise it some more.

"That little jig is an essential part of lifting the door!" remarked Ffennell. We must have looked at him strangely, because he smiled.

"The farmer's body, the door, and the ground form an enclosed space in the shape of a triangle, right?"

"Right."

"And if he tries to expand that triangle, he's creating a partial vacuum, right?"

We nodded, beginning to get it.

"So how can he work against a vacuum? In effect, it gets to be like lifting the entire weight of the Ardean atmosphere. The little jig is just to equalize the pressure under the door."

Bravo. In one stroke, Ffennell had not only shown that Yendred was indeed behaving himself but had illustrated once again the need for caution when interpreting Ardean affairs.

When the door was at last in place, the farmer and his wife took Yendred across it and into their house, where they brewed a hot drink of some kind, followed by a sort of soup. Eavesdropping, we discovered why the old couple lived alone.

WHY YOU AT HOME NO SONS OR DAUGHTERS HAVE?

THEY TO THE CITY HAVE GONE. WE THIS LAND MUST WORK ALONE.

YOU HOW OLD ARE?

I 107 AM AND HE 110 IS. WE TWO SONS HAVE BUT THEY TO FARM DO NOT WISH. ONE A ROCKET PILOT IS. THE OTHER IN A GLUE FACTORY WORKS.

The two invited Yendred to stay over with them until next morning, but Yendred was clearly anxious to be off. After warning him about something called a "Ra Nifid" east of the farm, the farmer and his wife said farewell to Yendred as he climbed the swing stairs. They sat for a time, quietly sipping more of the hot brew.

About two kilometers east of the farm, Yendred came to a hole in the ground beside which lay a stone slab. On the other side of the hole was a small mound of soil with a shallow pond behind it extending some twenty-five meters eastward.

From our vantage point, we could see more of the hole than Yendred could. In reality it was a vertical burrow with a single short, horizontal chamber in which a horrible-looking creature quietly waited.

▪ YNDRD. BE CAREFUL. THERE IS A STRANGE ANIMAL DOWN THAT HOLE.

IS IT MOVING OR STILL?

▪ IT IS NOT MOVING RIGHT NOW. WHAT IS IT?

THIS IS THE RA NIFID WHICH THE FARMERS WARNED ME ABOUT. THEY CAN BE EXTREMELY DANGEROUS. I MUST BE CAREFUL.

▪ WHAT WILL YOU DO?

AM I STILL SPEAKING WITH ALICE?

▪ YES. AREN'T YOU AFRAID?

NO. IT IS JUST A MATTER OF MOVING VERY SLOWLY AND QUIETLY.

Yendred stood for a time, considering what to do. Then he moved slowly away from the hole, retracing his steps. Stopping, he released the balloon kit from his lower arm, found the two ends of the balloon, and clipped them into place over the mouth of his gas bottle. Clearly, he planned to fly over the Ra Nifid burrow.

▪ WATCH OUT. IT IS STARTING TO MOVE. NOW IT IS CLIMBING UP THE BURROW SLOWLY. CAN YOU HEAR IT?

NO. THE GAS IS TOO LOUD. IS IT NEAR THE TOP?

▪ YES. HURRY. HURRY.

The balloon was now half inflated. It bobbled over Yendred's head, dangling the bundle of his belongings. The Ra Nifid emerged from its hole and paused to survey the intruder on its domain. The balloon was larger now but not yet light enough to hoist Yendred out of danger.

I CAN SEE IT NOW. I THINK IT IS GOING TO ATTACK ME.

Sure enough, the Ra Nifid made a sudden dash at Yendred, its three legs surprisingly nimble and well coordinated. Yendred hopped away from it, each hop carrying him farther as the balloon's lifting power became equal to Yendred's weight. But the Ra Nifid caught hold of one of Yendred's legs in its claws, just as Yendred began to soar aloft.

A curious situation now developed: the Ra Nifid, standing erect on its two rear legs, held Yendred's foot in its claws while Yendred, kicking furiously, held on to the balloon, which by now had too much lifting power for the Ra Nifid to overcome as it scrabbled on Yendred's leg, trying to bring him to the ground.

The situation might have remained static for quite some time were the balloon not steadily inflating. In less than a minute, it had become light enough to hoist both Yendred and the Ra Nifid aloft—which it did. The afternoon westerly was nearing its peak. The balloon, Yendred, and the Ra Nifid, floating heavenward, were immediately caught by the wind and soared over the Ra Nifid's hole and the little hill of soil. Frightened, the beast let go of Yendred's leg and fell with a splash into the pond. Immediately, Yendred shot almost straight up and off our screen.

"Lambert, take the terminal!"

Alice's decision had been instantaneous. We would need someone with real dexterity at the keyboard. Lambert's prowess at computer games was legendary in the department. This one could be called "Find the Balloon," or, better still, "Find the Balloon Before the Earth/Arde Link Dissolves."

Frantically, Lambert typed in different search vectors and tracking speeds in order to relocate Yendred. He finally found him by increasing the scale factor until the surface of Arde reappeared at the bottom of the screen with a miniature burrow and a tiny Ra Nifid wading out of its pond. Near the top of the screen was the balloon, still rising quickly, with Yendred dangling from it. If anything, the situation as now more serious than ever.

▪ HANG ON. DON'T LET GO. CAN YOU LET OUT SOME GAS?

THAT IS WHAT I AM TRYING TO DO.

Yendred could not merely detach one of the balloon's ends and let it go, for the balloon would then collapse totally and he would plummet to the ground. But he did detach one end with a lower hand and pass it to the upper hand. In this way, he was able to release whatever gas had made its way into the space between his arms.

AM I STILL GOING UP?

▪ YES. LET OUT SOME MORE GAS.

He repeated this maneuver several times until his ascent slowed. Then he began to drop and, panicking, started up the canister. Lambert's cool head prevailed, however.

▪ DO NOT REFILL THE BALLOON. STOP IT. YOU HAVE A LONG WAY TO GO. DON'T WORRY.

MY ARMS ARE VERY TIRED.

▪ THAT IS WHY WE MUST GET YOU DOWN QUICKLY.

It was a matter of astute judgment when to instruct Yendred to begin refilling the balloon. But Lambert found it quite easy.

"It's just Moon Lander," he remarked casually.

"Moon Lander" is a very primitive computer game which we sometimes run as a demonstration for visitors during our annual open house. A lunar module descends under gravity toward the moon's surface and the player must judge what rocket thrust to use for the gentlest descent before the fuel runs out. This is precisely what Lambert did for Yendred. He finally landed on the Ardean surface with a slight jar and immediately let go of the balloon, which collapsed at his feet. He flapped his arms and worked his fingers scissors-like, as though they were sore.

```
THANK YOU. THANK YOU ALICE.

▪ LAMBERT HERE. DON'T MENTION IT.

THANK YOU LAMBERT. WHY SHOULDN'T I MENTION IT?
```

Lambert looked very pleased with himself.

Yendred, exhausted, made himself a seat on the ground and pulled the balloon skin over his western side to act as a windbreak. When we felt that he had recovered sufficiently from his adventure, we asked him about the Ra Nifid.

What we had seen was a typical Ra Nifid hunting burrow, which may extend up to a hundred meters in depth where the soil allows it. The Ra Nifid waits quietly in a chamber near the top of the burrow, listening for the telltale scrabbling noise of some burrowing creature accidentally breaking into it. The bottom of the hole is filled with water become foul from the discarded remnants of previous prey.

My memory of this scene is so vivid that in imagination I have sometimes felt myself to be a disembodied presence in a Ra Nifid's burrow, watching the dull glint of its smooth bones in the obscurity of the waiting chamber. The Ra Nifid's head is cocked expectantly downward. The air of the burrow is filled with a faint stench of rotting flesh; there are no breezes or currents in that still place.

Suddenly, the echo of small bits of soil plopping into the water far below reaches the Ra Nifid's ears. It pauses only for a moment and then propels itself forward like some well-oiled machine to fall, bottom first, into its own bottomless hole. It skitters down the hunting burrow, leg by claw, watching and feeling for the telltale disturbance in its smooth surface, coming at last to a crack in the soil behind which a small armored, slithering thing works desperately and clumsily backward from the only mistake it has made in its life. Powerful claws tear at the soil above and below the hapless burrower until its head is exposed, chittering with fear, legs working a minute defense.

Plucked from the soil, it is torn into pieces and held in four separate claws. Blood running down the Ra Nifid's arms and collecting in pools at the joints, the lunch is taken at leisure.

Only when it is satisfied does the Ra Nifid clamber slowly up to its chamber, bits and pieces of its victim sliding from its bones and dropping into the pool below.

Besides the foregoing method of obtaining food, the Ra Nifid constructs a dam on the upslope (eastward in Punizla) side of its burrow, in order to trap enough water to create a pond. A kind of algae grows in great quantities in such ponds, especially when encouraged by the presence of the Ra Nifid's wastes. Periodically, the Ra Nifid grazes upon these to supplement its diet of underground flesh.

The Ra Nifid is an intelligent creature, bearing, it would seem, a high degree of kinship to the Ardeans themselves, differing only in the structure of its hands, the shape of its body, and its habit of walking "on all threes" as it were. Can we suppose that the Ardeans are descended from such a creature? Given the lack of any other candidates, this seems a reasonable idea. But what of evolution on Arde generally? Has all life on Arde unfolded in tiny steps as things here on Earth have apparently done?

Being an intelligent creature, the Ra Nifid is susceptible to a form of companionship with the Ardeans, who have a special fondness for a smaller species of this animal.

It is perhaps reasonable to include here, on account of this same intelligence, a brief point about two-dimensional brain tissue. It was supposed many years ago by someone on Earth who speculated in this direction that two-dimensional brains are impossible because of the tendency for each nerve to block all others from crossing it. Each circle of nerves would cut off any process inside it from communication with the larger neural world. We too had begun to wonder how intelligence was possible in two dimensions (in spite of the evidence before our eyes), until Yendred visited a kind of Punizlan scientific academy where we learned, indirectly, that fully a quarter of the Ra Nifid's brain, and up to a third of the Ardean's, is taken up by small triads of cells which allow neural signals to cross over each other. (See the Appendix.) In this way the same level of interconnection required by brains like ours is available to the Ardeans.

We left Yendred sitting on the windswept Ardean surface, sheltered in his balloon skin and nursing another of his "brain aches." The next contact time was arranged and we terminated the 2DWORLD program.

The next contact was to take place the following evening, Earth time. But when I got into my office that afternoon, I hardly had time to shuffle the papers on my desk and settle back with my wake-up coffee when Chan poked his head around my door.

"Have you heard, Dr. Dewdney?"

"Heard what?"

"The DEC 10 is down. Looks pretty serious."

I jumped up and followed Chan into the lab, where some of our systems people were going over the machine.

"What happened?"

Laura McReady, the systems manager, fixed me with her green eyes. "Somebody's been horsing around with this thing!"

Just a few days after a thorough maintenance check, three separate kinds of hardware faults all showed up at once. So far she had located a number of bad disk areas, several faults in main memory, and a failed PC register buffer. I wondered how one person could do all that.

"Oh, easy," said McReady. "Open the disk pack and blow in some smoke, get a six-volt battery with two leads and send random currents into the memory chips, and so on."

In short, Laura suspected sabotage, presumably by someone with access to the computer laboratory. She took these suspicions to the department chairman, who immediately confiscated all keys and ordered the locks on the laboratory doors to be changed.

This unhappy state of affairs put an immediate end to the universe/Planiverse link. Alice fretted about Yendred, who at this very moment may have been facing unknown dangers on the barren reaches of Punizla. Edwards began to explore the possibility of running the 2DWORLD program on some other computer with a similar graphics display unit.

Applications for new keys had to be filed by Little and Edwards, both of whom had legitimate projects on the computer; we would at least have access to the laboratory when the DEC 10 became operational once again. In the meantime, I tried to prepare both myself and the students against the possibility that contact might never be renewed. I warned them that we knew too little about the factors making that contact possible. One little change in the hardware might throw an exceedingly delicate equilibrium of software and hardware out of balance.

For several days after the sabotage incident, I worked long hours in my office and at home, trying to interest myself once again in the analysis of data structures, a research field which I had neglected for too long. But things were not the same. I used to have the capacity to work on a problem in my mind for hours at a stretch. In fact, my wife would often come upon my inert form at such times and suspect me of chronic laziness. But I was in another world, a universe of abstract forms which I was trying to manipulate toward the solution of a problem. Now, however, this universe kept retreating before the larger reality which had confronted me. A perfectly real universe, one entirely different from ours, had temporarily revealed itself. I could not concentrate on my research problems.

It was no longer a happily prone body which my wife found at home during these days, but a nervous, pacing one. Five days into the enforced lull in Earth/Arde communications, she approached me, a sympathetic look in her eyes.

"Why don't we go to the island?"

I thought of our secluded island retreat in the midst of Canada's northern forests. The sun would be setting on the cabin just now and soon the loon's rippling call would echo mysteriously across the dark waters and resound upward into the glowing canopy of the universe itself.

Within two days, preparations for the trip were complete. The systems people assured me that it would be at least two weeks before replacement parts for the computer arrived. On the evening of Monday, June 16, I held a brief meeting with the students, and on the next morning we left, my wife, son, and I, for the lake. The drive took two days.

It was therefore on the evening of June 19 that I found myself staring gloomily at the very sunset I had earlier promised myself. My wife sensed problems.

"I've never seen you this moody. Never!"

I toyed with the idea of telling her.

"It has something to do with the university, doesn't it? With Alice?" she added helpfully.

"Indirectly. I can't work properly anymore. I can't sleep."

A freshening breeze rippled the waters in front of the cabin, and I heard myself begin to speak. The strange messages over the printer in May. The view of Arde and its inhabitants on our display screen. By the time I had finished, the stars were out and loons were laughing in the distance.

"Are you making this up?"

"I almost wish I were. No. I'm afraid I'm not making it up. That's the whole problem, really. That's what's eating away at me. On the one hand I have this incredible opportunity and—responsibility. On the other hand, how is anyone going to believe me, even with the transcripts and photographs and videotapes? *You* believe me, don't you?"

"Let's go inside. I don't like it out here." She glanced apprehensively up at the impossibly remote distances. "Why shouldn't anyone believe you?"

"Because it's so insane, so bizarre, so utterly improbable."

We entered the cabin to find my son grinning at us. "I heard every word you said out there."

"Well? What do you think?"

"Sounds pretty real—unless you're completely insane!"

My wife glanced at him sharply and the boy fell silent. A chill passed through me. It was then I realized that, indeed, insanity could be exactly like that: one could imagine anything and imagine it so vividly that it became real. There really was no way of knowing.

"Anyway, we're here on vacation, and I'm getting cold. Let's build a fire."

I actually succeeded, toward the end of our ten days at the lake, in getting back into a research problem. I would hedge my bets.

The day after our return, the new parts arrived by courier and within hours they had been installed and tested. We then only had to wait our turn to use the computer.

Friday, July 4, 2:00 P.M.

The whole group was there. Edwards set up the 2DWORLD program and I took over at the keyboard.

```
▪ RUN 2DWORLD.
```

The landscape of 2DWORLD appeared on the screen. An FEC stood in the middle of it.

▪ HELLO YNDRD.

UNKNOWN: ''HELLO.''

"Wait, Dr. Dewdney, wait. We haven't got the link yet."

Then it happened, that magical transformation in which straight lines become crooked and simple geometric organs take on alien convolutions.

▪ HELLO YNDRD.

WHERE HAVE YOU BEEN? YOU HAVE NOT SPOKEN FOR 17 DAYS AND I AM ALMOST AT MAJ NUNBLT.

Yendred said all this as he floated peacefully along, suspended from his travel balloon. He came to a gentle landing and then, pushing off, floated up and to the east once more.

▪ OUR COMPUTER [erase line] OUR COMMUNICATION MACHINE HAD AN ACCIDENT. IT TOOK 21 OF OUR DAYS TO FIX IT. TELL US PLEASE WHAT HAS BEEN HAPPENING TO YOU.

Yendred had expected to be at Maj Nunblt much earlier than this but had lost time due to an unfortunate circumstance: camping out one evening, he had been surprised by a river of unusual size and speed. Unable to find a house quickly in that sparsely inhabited region, he had barely time to inflate his travel balloon a little. Hardly sufficient to lift him out of harm's way, the partially inflated balloon nevertheless buoyed him up in the river quite well. Wedging it between the two arms on one side, he was able to lie upon the balloon and hold his other two arms in the air in order to breathe. Ardean gills do not operate at all well in water.

The river carried him about seventy-five kilometers by Yendred's reckoning, gradually subsiding in an enormous slough from which his only escape was by inflating his balloon to capacity. By then, the pre-dawn easterly was in full force, and he was carried another thirty kilometers.

First, venting the balloon, he had fallen back into the quicksand, from which he escaped once again by reinflating—more judiciously this time. Terrified, cold, wet, and tired, he went up and down, alternately bounding, being dragged by the balloon through the mud, and revolving helter-skelter in the turbulence of that awful wind.

When at last he came to dry ground, he deflated the balloon and sought shelter in the first house he encountered. Depressed and exhausted, Yendred had fallen ill and the family had nursed him. But not even the presence of little children wanting to hear his adventures could cheer him up.

Yendred interrupted his narration to inform us that this was the last time he would be able to use the balloon; his cylinder was nearly empty and he had discarded it. Yes, it was clearly missing from the pack of belongings which hung from the balloon. The city of Maj Nunblt was yet another two days' travel, and the afternoon westerly was already a spent force.

Descending a last time, Yendred deflated the balloon and placed the skin in his pack, which he now carried. As he walked, he continued to talk about events of the last few weeks and then suddenly asked us,

EARLIER YOU SAID ''COMPUTER.'' WHAT IS IT?

▪ A MACHINE WHICH CALCULATES. ACTUALLY IT DOES MUCH MORE THAN THAT. IT CAN HELP US ANALYZE DATA, MAKE DECISIONS, PREDICT THE FUTURE, AND DESIGN THINGS. IT CAN EVEN PLAY GAMES AND TALK TO US. SOME HUMANS THINK THAT COMPUTERS WILL EVENTUALLY BE AS INTELLIGENT AS WE ARE.

YOU TALK TO ME WITH THE COMPUTER?

▪ YES. BUT WE HAVE NO IDEA HOW IT WORKS.

BUT YOU BUILT IT.

▪ WE BUILT THE COMPUTER BUT WE HAVE NO IDEA HOW IT ACTS AS A COMMUNICATION MACHINE. IT IS VERY MYSTERIOUS.

WE HAVE COMPUTERS I THINK.

This was a surprising statement indeed. So far, what we had seen of

Punizlan technology would place it in the late nineteenth century by Earth standards. The "computers" that Yendred next described were experimental machines developed at a famous center for science and technology in the very city which he now approached. Apparently, our views of Ardean technology were in for a drastic revision!

The center was really rather close to what we on Earth would call an "institute." The best and brightest of Punizlan scientific minds were gathered there, spending most of their time in basic research or technological development. The institute also acted as a mini-university, taking some dozen new students every year. The students learned by assisting the members of the institute with experiments. Every morning, the members and students would gather for a seminar. This was as close as they ever came to formal lectures. Yendred knew a great deal about the institute because he once had ambitions of attending and had eagerly learned everything he could about it while attending his last year of senior school (something like a combination of late secondary school and early university). But he had done rather poorly in his last year, for at this time he had first come under the influence of ideas which were to result in his present trip. He had spent his time dreaming rather than studying. This much Yendred freely admitted.

Saturday, July 5, 2:00 **P.M.**

The next day saw Yendred plodding onward toward the city of Maj Nunblt, small by comparison with the Punizlan capital and giving, as yet, no sign of its existence by an increase in the number of houses underfoot. Indeed, the landscape seemed just as barren as ever.

Around midafternoon, Yendred heard a river. Having passed no houses recently, he wobbled ahead as quickly as his short legs would carry him. The river was coming.

IS THERE A HOUSE EAST OF ME?

We scanned rapidly east on the screen and came, within fifty meters, to a flat stone covering a narrow vertical shaft. Ra Nifids use just such stones to protect their burrows from water. Scanning down the shaft, we saw the characteristic waiting chamber, but no creature in it. The burrow was clear right to the bottom.

▪ YES. THERE IS AN ABANDONED RA NIFID BURROW. KEEP GOING.

By the time Yendred reached the flat stone, the river was already there. It seemed to be only a few centimeters deep, hardly something to take refuge from. Yet sometimes such rivers deepened with time, Yendred told us, the crest of the flood coming near the end. He lifted the stone, jammed himself into the burrow, and worked his way down, pulling the stone overhead as he went.

ARE YOU SURE THERE IS NO RA NIFID HERE?

▪ ABSOLUTELY.

He made his way down to the chamber and suddenly froze.

THERE'S SOMETHING HERE.

We scanned to the end of the chamber and received a very nasty jolt. A pair of arms appeared, and then a body. Thank God, it was an Ardean body.

▪ YES. NOT A RA NIFID. NOT A RA NIFID. IT IS AN NSANA.

The Ardean must have entered the abandoned burrow while we scanned back to Yendred. Now the two exchanged greetings in the dark, smelly Ra Nifid chamber. The stranger was called Tba Kryd. He too was traveling, but in the opposite direction, having left Maj Nunblt the evening before. He described himself to Yendred as a philosopher, one who tried to learn about the meaning of existence. He turned on a small hrabx canister near his pack and continued.

WE ABOUT OURSELVES LEARN MUST. WE OUR BRAINS AND BODIES EXAMINE MUST. WE OUR TECHNICAL PROGRESS LIKE BLIND BURROWERS TO DEVELOP CANNOT CONTINUE. OUR PURPOSE ACCIDENTAL NOT IS BUT TO OURSELVES BY OURSELVES IS GIVEN. WITH SUCH KNOWLEDGE OF OURSELVES WE OURSELVES MAY IMPROVE.

YOU A SCIENTIST ARE?

I A SCIENTIST WAS BUT NOW A PHILOSOPHER AM. PUNIZLA LIKE A GREAT ROCKET SHIP IS. MY MISSION ITS DIRECTION TO CHANGE IS.

OF THE PRESENCE WHAT?

THE PRESENCE AN IDEA ONLY IS.

OF THE SPACE BEYOND WHAT?

THE SPACE BEYOND MERELY TIME IS.

TIME AND ETERNITY

The philosopher's statement makes sense if we interpret the "space beyond" as a higher dimension. In such a case, his statement is quite consistent with the view of modern physics that time is an extra dimension. Thus we live in a universe having three spatial dimensions along with the additional dimension of time.

To see this, imagine a succession of instants at a particular place on Arde. For example, imagine two disks rolling toward each other, colliding, and then rebounding. If we think of successive instants as frames in a movie film and if we then stack up all the frames, in order, an interesting picture develops,

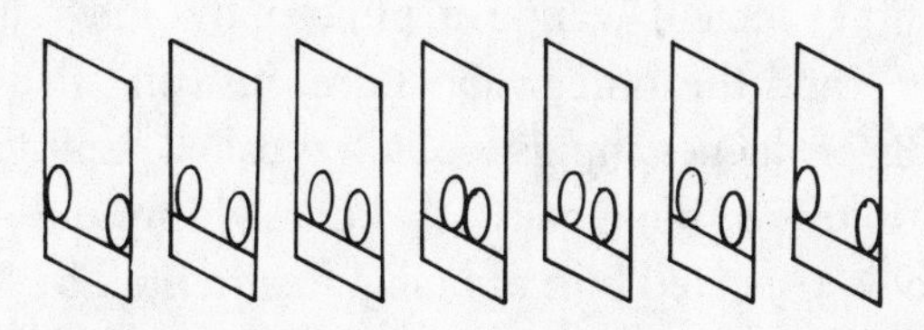

especially if we fill in all the gaps, so to speak, and discover that a new, three-dimensional shape emerges.

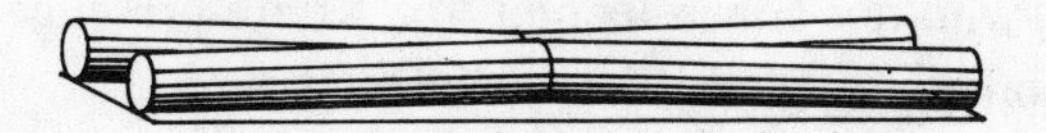

Here is a "timeless" picture of the colliding disks. Better still, it is a picture of the two disks in eternity, an eternity in which all instants of time lie frozen in place.

Although a trifle too sure of himself, I found Tba Kryd a rather interesting figure, one who expounded a challenging philosophy as well. It occurred to me that we might be watching one who would have an important influence on the development of Arde in the future, and I was curious about what influence he would have on Yendred now.

> YOU TO VANIZLA HAVE BEEN?
>
> I TO VANIZLA HAVE BEEN AND I THE VANIZLAN SAGES HAVE MET. I THEIR TEACHING HAVE ABSORBED AND I THEIR RELIGION TO THE DEPTHS HAVE EXPLORED.
>
> [There was a long pause before Yendred's next question.]
>
> YOU WHAT DISCOVERED?
>
> THEY INTERESTING IDEAS HAVE BUT THEIR KNOWLEDGE SUBJECTIVE IS. IT SCIENCE NOT IS AND NOTHING UPON IT BE BUILT CAN. ONCE ALL ARDE LIKE VANIZLA WAS BUT VANIZLANS NOW BEHIND LEFT ARE.

Yendred asked no more questions, and Tba Kryd, taking this for an expression of interest, gladly filled the silence with a long description of his travels in the land of Vanizla and what he had learned there. Two subcultures dominate the affairs of Vanizla. One of these, clinging to the outer forms of the ancient Vanizlan philosophy and religion, could be called "legalists," and the other subculture, insisting that external forms are only symbols of deeper things, could be called "spiritists."

Tba Kryd told how the legalists had chased him out of a town called Flus because they suspected him of being an atheist. Tba Kryd spoke with apparent bitterness of their lack of tolerance and their insistence that the whole population observe the rituals of public and private behavior to the letter. From this hostile presence he was driven into the arms, so to speak, of the spiritists, who not only tolerated Tba Kryd's presence but attempted to teach him something about the Vanizlan system.

External forms, including material objects, are but the symbols of deeper things, manifestations of a thing called "the Presence." All things point to the Presence, more or less, and have "real value" only by how directly they so point. The spiritists go on to claim that the Presence may be known directly, rather than merely reflected upon, by ascending a chain of experiences to the Presence itself. This claim is hotly disputed by the legalists, who call it a mere fancy and a pretension aimed at gaining influence for the spiritists among the Vanizlan population.

Although Tba Kryd was never able to absorb the supposed reality of the spiritist teaching, including the chance to experience a "death before death," he had probed tirelessly for any substantive knowledge which the Vanizlans might have. To his surprise, he had discovered that many of the machines of which the Punizlans were so proud had first been built long ago in Vanizla. But these ancient machines had never been fully exploited, as though their builders regarded them as curiosities or in some way beneath their dignity to develop further. This attitude greatly perplexed Tba Kryd, and mingled with his admiration for these Vanizlan inventors was a certain annoyance at their lack of imagination.

Tba Kryd turned next to Punizla, how in ancient times it had been very similar to Vanizla. In particular, he told Yendred the tale of a very strange Punizlan named Ajemsana who "the world constructed and mad went." He spent his whole life speculating on the nature of things, finding in each appearance the cause of that appearance. Treating each cause as a new appearance and looking into the causes of these, he at last reached a state where he would talk to no one. Before reaching this state, he made a number of cryptic utterances, including "The universe itself does not support," meaning, one supposes, that the universe does not support itself, whatever *that* means.

Tba Kryd at last finished talking, and Yendred asked no further questions.

▪ ASK TBA KRYD HOW HE PLANS TO GUIDE THE AFFAIRS OF ARDE.

I WOULD RATHER NOT.

▪ WHAT IS WRONG?

The long and short of it was that Yendred had become very depressed and unhappy after hearing Tba Kryd's long description of the Vanizlans and the rather limited extent of their knowledge. He wondered whether it was worth going on. Here again, we gave him no advice but by degrees got the conversation back to Tba Kryd. Would Yendred ask him about his plans, as a favor to us?

At last Yendred complied, shrugging his whole body as though sighing.

Tba Kryd's reply came immediately. He had plans for an entirely new form of government geared to a specific program by which Ardeans might reach the very peak of personal and public perfection. Their science and technology would leap ahead. Their social organization would be perfectly geared to the happiest possible life for all, and Planners and Naturals would exist no longer.

Yendred climbed up the Ra Nifid burrow to remove the stone, and Tba Kryd turned off the hrabx bottle. They settled down to sleep and we dissolved the Earth/Arde link.

7

The Punizlan Institute

The city (or should I say "town") to which Yendred came two days after his meeting with the philosopher was called Maj Nunblt, a place inhabited by barely a thousand souls and entirely dominated by a sort of institute which we had heard about earlier. Because of its similarity to research institutes on Earth, we decided to call it the Punizlan Institute of Technology and Science. Working at the institute and living nearby were some fifty scientists and technologists, supported by a large number of technical staff who helped with experiments or built equipment.

The Punizlan Institute dominated the central portion of Maj Nunblt and consisted of eight separate underground buildings, one for each of the major sciences:

Ardesystems	Reactionsystems
Largesystems	Smallsystems
Lifesystems	Thoughtsystems
Machinesystems	Wavesystems

Each building contained laboratories, work areas, and offices. It would be no exaggeration to say that what we found here boggled our minds. Personally, I didn't know whether to be more impressed by *what* the Punizlans had discovered or the mere fact that *they* had discovered it. Who, after all, can admire the universal law of gravitation without admiring Newton?

Yendred's guide in Maj Nunblt was a former schoolmate, Ladrd by name, who was now an apprentice scientist at the institute. He showed Yendred around on two successive days and we, his invisible partners, followed the tour with mounting fascination. Our tour, barely begun, was suddenly interrupted by events on Earth. Unwanted publicity was about to hinder our daily contact.

Both before and after this unwanted interruption, we kept Yendred so busy asking questions that his hosts may have become a little annoyed with him. I prefer to think, however, that true scientists, in whatever universe they dwell, are pleased to inform inquiring minds.

Wednesday, July 9, 2:00 P.M.

The tour began in the Wavesystems building, where Yendred was introduced to a sort of applied mathematician named Tba Bindl whose specialty was the analysis of waves, both in the abstract and in nature. Tba Bindl had not only improved the wave equations recently developed in Punizla but had solved them in a certain special case which explained a phenomenon of Ardean sound that had long puzzled Punizlan scientists.

Every time an Ardean speaks or a tone is played or a gong struck, indeed every time any sound is heard on Arde, the frequency edges up a few semi-tones during the first second of its being heard. This briefest of rising glissandos precedes all sounds, and the effect is more pronounced at greater distances.

Tba Bindl's explanation of this effect, as due to the basic nature of the Planiverse, had already earned him great fame. The same effect, he claimed, holds for light and all other forms of wave-propagated energy. The shift in light frequency takes place far too rapidly for the Ardean eye to detect, but the shift was recently measured by instruments designed for the purpose, a triumph for Tba Bindl's theory.

This young but already accomplished scientist explained to Yendred and Ladrd how he hoped to explain another phenomenon of Ardean sound: certain wave forms turn into others at a distance. What might sound nearby like one musical instrument sounds like quite another at a distance. Another solution of the wave equation had already predicted how a square wave would turn into a sawtooth wave at a distance.

Even before the tour and Yendred's visit to Maj Nunblt, the students and I had speculated on Ardean sound and had realized that sound, light, gravity, any and all forms of energy propagated by waves, were much longer-lived in the Planiverse than in our universe. This is entirely due to the differing dimensions of the two spaces: sound and light, like gravity, diminish by the rule $1/d$ (d being distance from the source) in the Planiverse, while in the universe they diminish much more quickly—by the rule $1/d^2$.

DIMINISHING ENERGY

The reason that energy in our universe diminishes according to the formula $1/d^2$ is really quite easy to understand. Imagine, for example, that a source of light shines upon a square surface at a distance of 1 meter and observe that at d meters the same light would illuminate a larger square consisting of d^2 of the smaller ones.

For example, at a distance of 3 meters the larger square would consist of $3^2 = 9$ little squares, each of which would receive only 1/9 of the light available at 1 meter. By this reasoning, it seems apparent that light (and any other form of energy propagated along straight lines) must diminish to $1/d^2$ of its strength at 1 meter when it is observed d meters away in our universe.

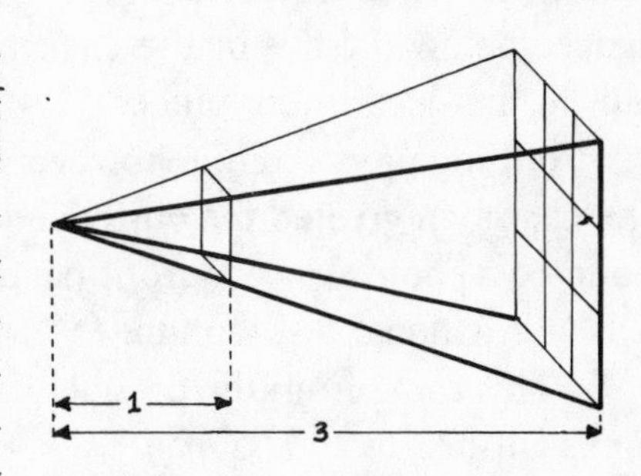

In the Planiverse, only the bold lines in the diagram above are relevant. A light shining on a line segment 1 meter away illuminates a line d times as long d meters away, and each of the d line segments composing it receive only $1/d$ as much light as the segment 1 meter away. That, in a nutshell, is the Planiversal rule governing the rate at which sound, light, and gravity diminish with distance from the source.

Apart from whatever difficulties are produced by the anomalies of Ardean sound waves, there can be little doubt that, other things being equal, Ardeans hear each other over much greater distances than we can. Indeed, we did once observe Yendred cock his head, apparently listening to a conversation taking place nearly eighty meters away on the Ardean surface. On the other hand, Ardeans do not hear each other nearly so well in enclosed spaces, and we think that the reason for this is that very same relative strength of sound with distance: echoes develop much more easily and are sustained longer.

As Yendred and Ladrd descended the stairs to another laboratory in the Wavesystems building, something of greater importance suddenly developed in our own laboratory. One of the department secretaries entered the laboratory and pointed her finger at me. The students were frantically trying to get the Star Trek game up on the computer, but she seemed not to notice.

"The chairman wants to see you."

I could tell from the undertone of delighted menace in her voice that something was wrong. Clearly, any professional academic in my position would have developed guilt feelings by now. I had been neglecting my research and had missed a number of important committee meetings. I dreaded being told this by the chairman, and made my way with reluctant steps to the department offices.

Nothing prepared me, however, for the reception which awaited me. The chairman greeted me curtly and with a pained expression on his face proffered a photocopy of a front-page story in a certain well-known tabloid with international distribution.

"This came from the president's office this morning. Would you like to explain it?"

Horror rose in my throat. The headline swam in my vision:

"PROFESSOR DISCOVERS FLAT WORLD. University Student Tells of Computer Contact With 2-D Beings."

"What, precisely, is going on around here?"

"Nothing, really. I think the students got a bit carried away with our simulation project last term. That's all."

"I presume you mean the 2DWORLD thing."

The chairman seemed most unhappy. I spoke quickly. "Exactly. Look, our students are under tremendous pressure. Besides the continuous weight of long and difficult assignments, they see half their classmates missing every year. I think some of them are bound to go a bit bonkers along the way. Who was it, anyway?"

"No name given." The chairman studied me suspiciously for a moment. Then he said, very softly, "Crikey, man, whatever you're doing, stop it!"

In short, he was asking me to stay away from the computer laboratory and to get back to legitimate research. Strangely enough, in spite of my embarrassment at the story, both on my own behalf and on the university's, I came out of his office feeling almost relieved. I knew what had to be done.

Back in the laboratory, 2DWORLD was again up and running, events on Yendred's world continuing to unfold. The students looked up from the screen. It showed a new room in the Wavesystems building. Yendred and his host were standing with two other Ardeans.

"What was it?"

"Some very unhealthy publicity has developed. We're going to night sessions."

I told the students about the tabloid story and we spent some time speculating on who might have leaked it to the press. Realizing that this was an unhealthy exercise which could only lower group morale, I cut the discussion short. From now on we would meet only at night. This meant that we could contact Yendred only when his waking hours corresponded to our nights, roughly two out of every three days. For now we continued the present daytime session, warning Yendred at the first opportunity.

Yendred and Ladrd were talking with another physicist called Tba Shrin. This one studied the Ardean equivalent of electricity and magnetism. She was accompanied by a young apprentice who listened attentively and said nothing. Putting certain questions to her through Yendred, we discovered that electricity in the Planiverse is much like our own but that magnetism has entirely different properties from ours. In the Planiverse, magnetic fields have no direction, no "lines of force" so to speak, and are incapable of attracting anything. Magnets were discovered only recently in Punizlan history, when their effects on electrons were observed.

In our universe, when an electron enters a magnetic field, it experiences a deflection at right angles to the lines of force and, if the field is uniform, the electron travels in a circle.

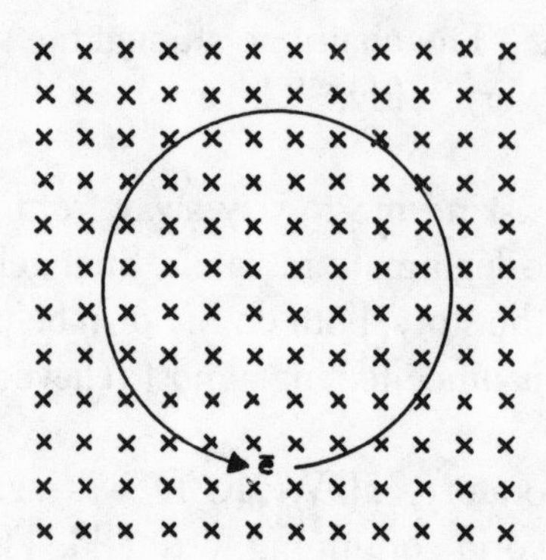

This is precisely the effect which Ardean magnets have on Planiversal electrons, as if the magnets were surrounded by points of force created by three-dimensional lines of force penetrating the Planiversal film. Ardean magnets come in two forms. One bends moving electrons into a clockwise path and the other into a counterclockwise path. In analogy to how Earth physics books represent lines of force pointing into or out of the page by the symbols "X" and "O," we named the two types of field accordingly. The intimate relationship between electricity and magnetism in the Planiverse was simultaneously comforting and eerie: in spite of the fact that each kind of phenomenon may be used to generate the other (as on Earth), the Planiversal magnetic fields were utterly different from ours.

Of special relevance to the use of electricity in the homes and industries of Arde is the presence of magnetic fields which enclose every current-carrying wire on Arde. Beside a very long wire, these fields do not lose strength with distance from the wire until the limits of two enormous triangular areas are reached. The wire forms the base of both triangles.

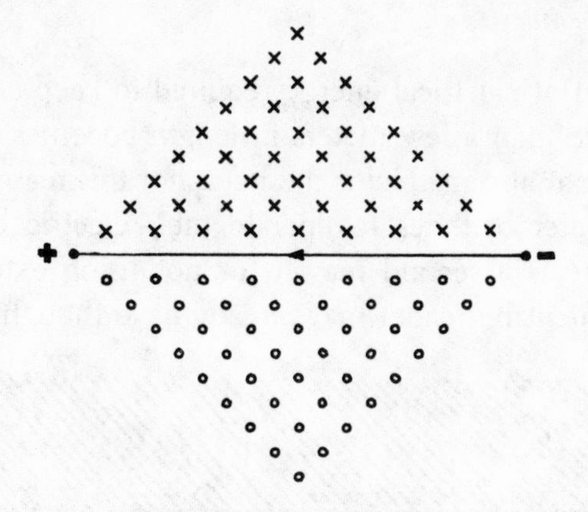

ELECTRICITY AND MAGNETISM

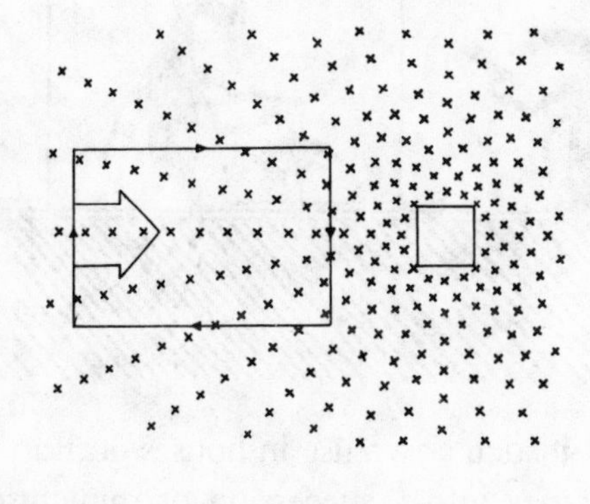

As a magnet and a loop of wire come closer together or move farther apart on Arde, an electric current is generated in the wire.

In an X-field, a movement of the loop toward the magnet creates a clockwise current in the wire, whereas a movement away from the magnet creates a counterclockwise current. In an O-field, the reverse happens.

Just as the relative motion of a magnetic field near a wire creates an electric current, so every electric current creates a magnetic field. More precisely, there will be an X-field on the "right" of the wire (from the current's point of view) and an O-field on the "left."

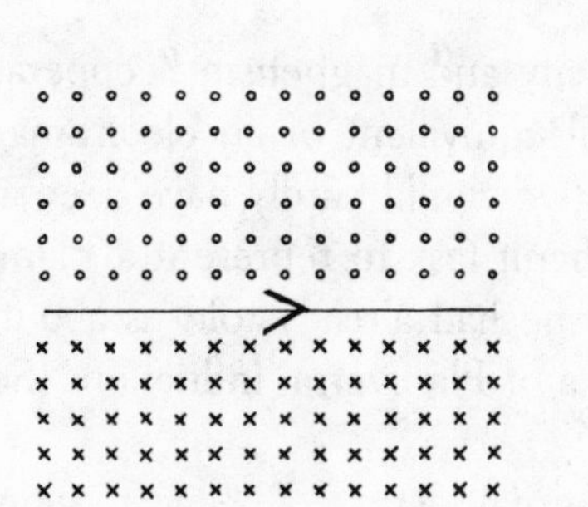

Thus, if the current in the wire is reversed, the two fields change position.

The amount of electrical energy required to keep a current flowing in a wire of more than a few meters in length becomes quite enormous because the current also maintains the field. For this reason alone, Punizlans certainly prefer batteries to operate their electric lights and other appliances. There is a second reason for not using extensive electrical circuitry, one which the reader may already have thought of.

Clearly, a standard parallel circuit such as we use in houses on Earth would compartmentalize Ardean homes into a succession of miniature prisons in which the beneficiaries of electrical living would suffocate in their own circuits!

Tba Shrin explained how electricity and magnetism "cooperate" in forming what is surely the Planiversal equivalent of an electromagnetic wave. Although she drew no pictures (we would hardly have been able to appreciate them anyway), it is no difficult task to represent a Planiversal electromagnetic wave based on what she had already told us about electricity, magnetism, and the phenomena of Planiversal induction. (See the Appendix.)

Near the end of Tba Shrin's informal lecture, I began to wonder if there was anything like the theory of relativity in Punizlan science.

▪ ASK HER IF LIGHT BENDS AS IT PASSES A MASSIVE OBJECT.

Tba Shrin's reaction to this question was a bit of a disappointment. She asked Yendred what could have caused him to ask such a silly question, and his friend Ladrd seemed to shuffle with embarrassment.

WHY DID YOU SUGGEST THAT I ASK SUCH A QUESTION?

■ ASK HER HOW A WAVE OF LIGHT WOULD APPEAR IF SHE COULD MOVE WITH IT AT THE SAME SPEED.

PLEASE. YOU ARE MAKING ME APPEAR FOOLISH. WHAT IS THE POINT OF SUCH QUESTIONS?

■ IF YOU ASK HER IT MAY BE TO HER ADVANTAGE.

Yendred asked Tba Shrin the question. She did not say anything for a long time. Her head swung into a vertical position as though she were imagining an electromagnetic wave passing overhead. Finally, she turned again to Yendred and asked him what *he* thought it would look like.

WHAT SHALL I SAY?

■ TELL HER THAT SUCH A THING IS IMPOSSIBLE TO IMAGINE.

Yendred, visibly trembling, did as he was asked, and Tba Shrin became rather animated. After dictating some notes to her student, she cocked her head to look at Yendred with both eyes.

THAT SUCH A THING IMPOSSIBLE TO IMAGINE SHOULD BE. THAT TO ME A LIMITING CONDITION ON THIS UNIVERSE SUGGESTS. AND UNRIPE THOUGHTS IN MY MIND APPEAR. YOU WHERE SUCH AN IDEA DID GET?

Yendred gave the Ardean equivalent of a shrug, weaving briefly from side to side, as though wishing to dodge the question.

LADRD THAT YOU A SCIENCE STUDENT ONCE WERE ME HAS TOLD. YOUR FUTURE PERHAPS HERE IS.

Throughout this interchange, Edwards had been looking at me strangely, his dark eyes brooding, his arms folded across his chest. He caught my eye. "Aren't you doing what you told us not to do?"

"What's that?"

"If I'm not mistaken, that's a gedanken experiment of the sort that leads to relativity theory."

I felt a blush of shame rise in my cheeks as I realized what I had done. I then remembered having the fantasy, only moments before, of making Yendred into a young Einstein, as though a mischievous child had taken possession of me. The thing I had done was truly monstrous, far worse than anything the students had done.

"You're right. You're right. I'm sorry."

Edwards studied his shoes.

Yendred, meanwhile, had replied that he would like very much to be a student at the institute. He told Tba Shrin that he had been on his way to Vanizla but had recently changed his mind.

> THEN YOU ME TOMORROW WILL VISIT AND WE OF YOUR STUDIES WILL SPEAK. I OF YOU MORE MUST KNOW AND YOU OF ME.

"Is this the end of his travels?"

No one answered Alice at first. She looked at me almost accusingly, and I felt compelled to speak.

"I don't think he was going to continue, anyway. His conversation with Tba Kryd took some of the wind out of his sails, I'm guessing."

Yendred and his host said goodbye to Tba Shrin and ascended to the surface on their way to the Smallsystems building. Here they met Tba Ftt, whose specialty was atomic structure. The actual phrase used by this scientist was "structure of the indivisibles."

The Punizlans had suspected the existence of atoms for centuries, but proof had come only twenty-five years ago. Subtle and ingenious experiments had confirmed the existence of the indivisibles, and further experiments, even more subtle and ingenious, had shown that even the indivisibles had a structure of their own and were, in fact, divisible.

In parallel with these experiments a kind of quantum theory had sprung up. As Tba Ftt reviewed the outlines of this theory to his young guest, I became excited at the parallels to our own quantum theory—what little I knew about it, at any rate. One difference that struck me immediately, however, was that Planiversal atoms appear to have only three quantum numbers, while those in our own universe have four.

After our contacts with the Planiverse ended, I pursued some of these matters with colleagues in physics. Using information which I supplied, one of them did some preliminary calculations and soon announced that almost any model he could develop for two-dimensional atoms leads to enormous sizes when compared with our own, in some cases ten or twenty times larger! Does this give us a kind of absolute yardstick from which we might conclude that Yendred, far from equaling my height, is ten to twenty times larger than me? The thought is bizarre and yet not out of keeping with all that we had so far found alien in the Planiverse. Nevertheless, I have gotten into the comfortable habit of regarding Yendred as exactly my height.

It was toward the end of this much briefer visit with Tba Ftt that Yendred thought of a question of his own to ask.

INDIVISIBLES OF WHAT MADE ARE?

SMALLER PARTICLES STILL.

AND THESE OF WHAT MADE ARE?

THE FUNDAMENTALS. NOTHING SMALLER IS.

THE FUNDAMENTALS OF WHAT SUBSTANCE IS?

OF ENERGY BUT INDIVISIBLE. A UNIT. A FUNDAMENTAL UNIT.

ENERGY OF WHAT MADE IS?

Ladrd began tugging gently on Yendred's arm. Yendred pulled his arm away.

ENERGY WHAT IS?

Finally, Tba Ftt replied.

ENERGY ENERGY IS. YOU WITH WORDS PLAY. ANYONE WITH WORDS PLAY MAY. YOU LIKE ONE RECENTLY OUT OF THE EGG ARE.

YOU LIKE ONE RECENTLY DEAD ARE.

We sat there at the terminal, stunned. We could hardly believe that Yendred, gentle Yendred, had said such a thing. Evidently, we had missed much of the underlying tension which had been rising between Yendred and the scientist. For one thing, we could not hear them directly. Would their voices have sounded a rising crescendo of squeaks, gabbles, and hisses? What signals of fear and anger could their impassive bony features send to each other?

▪ YNDRD. LEAVE HIM. APOLOGIZE AND LEAVE HIM.

I APOLOGIZE WILL NOT. HE THE MANNER OF A RA NIFID HAS.

Yendred allowed Ladrd to lead him away. We hoped that he had not spoken that last sentence out loud. In any case, some form of Ardean social unease made further touring of that building impossible, and the two clambered up a swing stair and out onto the surface.

The next building housed the department of Reactionsystems, a place where all kinds of chemical and physical reactions were studied. The upper portion of this building had a number of rooms with small heating units and liquid-filled vessels boiling upon them. In spite of the limitations of their two-dimensional environment, these scientists had designed an impressive array of ingenious-looking equipment. One of the scientists, Tba Nar, was doing basic research into the Planiversal elements.

Scientists on Earth have discovered over one hundred elements in our universe, but so far, the Punizlans have found fewer than fifty. Does this mean that the Planiverse has fewer elements? In writing this book I have done a great deal of reading in our own sciences, reacquainting myself with old subjects and painfully making my way through new ones. It seems that one reason for the relative stability of our atoms is that the nuclear particles (protons, neutrons, etc.) are attracted to each other by the "strong nuclear force" which acts at very short distances—the point being that in three dimensions, one can get a lot more particles within a small space than is the case with two dimensions. When the number of particles in one of our nuclei exceeds a certain limit, the strong nuclear force cannot act as effectively and the nucleus is unstable, becoming likely to undergo a spontaneous decay. This very instability may act at a much lower limit in the Planiverse. Presumably the heavier Planiversal elements also tend to be radioactive.

Adorning the ceiling of Tba Nar's laboratory was the Ardean equivalent of a periodic table of the elements. Tba Nar looked up at it frequently when explaining some of the problems he was currently engaged in. Yendred, appearing to listen attentively, began to recite the names of these elements and their periods. A few days later, it was a relatively easy matter to assemble our own version of the table. (See the Appendix.)

We did not even bother to learn the names of the Planiversal elements, they were so long. Nor did we have the time to find out much about Planiversal chemistry. However, there were elements which we could identify, by one means or another, as analogous to those in our own universe. For example, the simplest Planiversal element has a single electron and we decided to call this "hydrogen." The presence of a second electron fills the first Planiversal shell and the corresponding element is reasonably called "helium," as it reacts with nothing. The second shell may contain up to six electrons and we call the six corresponding Planiversal elements:

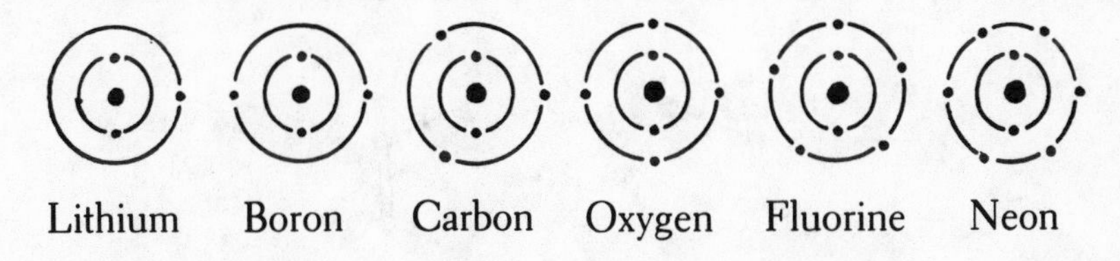

Lithium Boron Carbon Oxygen Fluorine Neon

In assigning these names we have been mindful of relative positions in the two tables. Thus, because carbon is almost midway across the first octet of our own table of the elements, I have given the same name to the corresponding element of the Planiversal table.

About all we could discover of Ardean chemistry was that their first element, "hydrogen," was plentiful both in the atmosphere and in the ocean. Presumably, it is a constituent of Ardean water.

Although this missed opportunity for knowledge depresses me somewhat, I am encouraged to think that our own scientists may be able to reconstruct at least some of the Ardean chemistry by the use of this table. As they do so, it will become increasingly obvious that the lack of a third dimension places a very strong restriction on what combinations and bonding structures are possible.

POSSIBLE AND IMPOSSIBLE MOLECULES

All Planiversal molecules, if drawn in the usual rod-and-ball symbolism, must lie in the plane. One can therefore immediately rule out any configurations which are not inherently planar. For example, no matter how one tries to arrange the molecule shown below, it simply will not fit into the plane: two lines always end up crossing, a thing which is presumably ruled out by the usual interpretation of the lines as bonds.

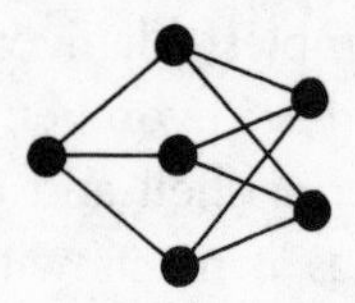

Given a molecule which does fit in the plane, however, there are almost always two versions of it, one the mirror image of the other.

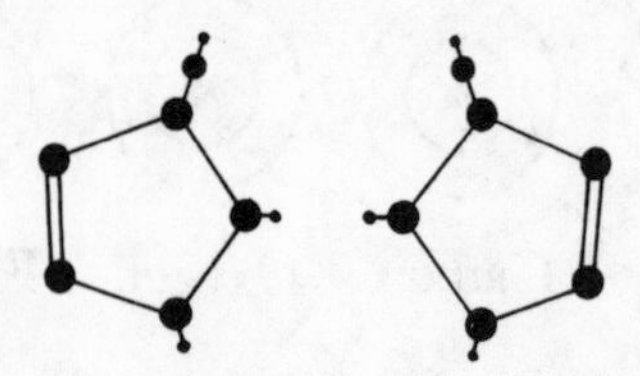

Neither compound may be converted into the other by any movement within the plane and, generally speaking, the two members of such a pair must have rather different chemical properties.

Finally, ionized atoms and molecules either cannot exist or are extremely rare in the Planiverse. The amount of work required to strip an electron away from a Planiversal atom or molecule is theoretically infinite. (This last speculation came to me after contact with the Planiverse ended and there was no way to confirm it by observation or questioning.)

During this visit to the Reactionsystems building, Yendred began to feel ill. Apparently the combined effect of all the ongoing experiments spewing their vapors into the building's stale atmosphere drove even the chemists to the surface occasionally for a "breath" of fresh air.

Thus it was that, having sampled only a small fraction of the Reactionsystems building's secrets, I watched regretfully as Yendred stumbled to the surface.

The next building housed Lifesystems, where scientists studied life in all its forms and manifestations. In the first room below the surface were some aquariums housing a variety of creatures that we hadn't seen before. Against the east wall, one particularly charming aquarium was filled with a number of very small, threadlike creatures called Zar Hyet. Besides these, algae-like organisms (Hadmsh Hab) floated near the surface and a sort of mushroom-shaped plant (Bar Prala) grew in the bottom.

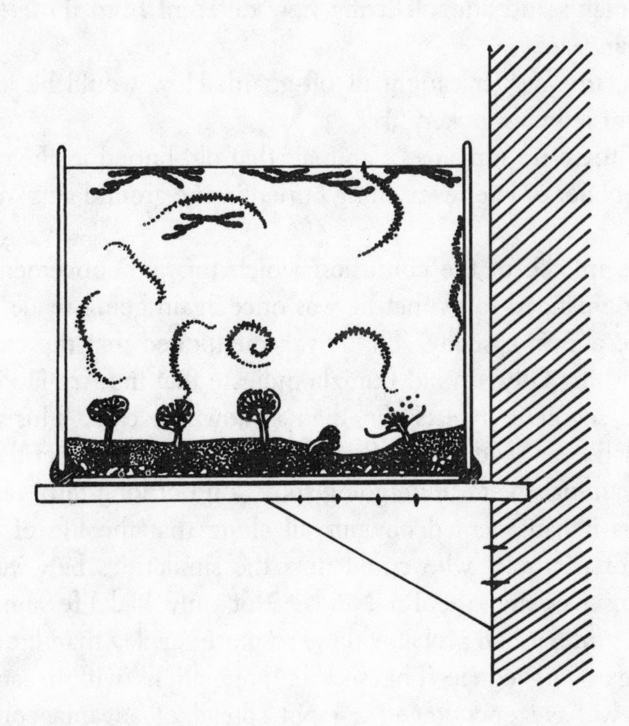

Most of the biologists at Lifesystems studied particular animals or plants, but the scientist Tba Hlyak seemed confined to no single species. It was through Tba Hlyak that we learned about life on Arde as a whole. Perhaps we should not have been surprised to learn that Arde has barely a thousand species of living creatures, plant and animal. The Earth has millions.

This vastly reduced diversity is no doubt due to the lack of living space on Arde, and this is meant not just metaphorically but in a rather literal sense as well: whether measured in cubic meters (as on Earth) or in square meters (as on Arde), a certain minimum amount of unoccupied space is required by each organism. On the other hand, each species requires a minimum number of individuals to guard itself from extinction. The more species there are, the more these requirements come into conflict, especially in Arde's limited environment.

Yendred's spirits seemed to pick up considerably in the Lifesystems building, and we easily pursuaded him to ask Tba Hlyak about evolution. Were the plants and animals living now different from those living millions of years ago?

Again, the answer caught us off-guard: How would he know what such ancient animals looked like?

Were there no remains of animals that old buried in the soil?

No. Sooner or later everything buried in the ground gets washed into the sea.

In the middle of the confusion which this announcement created, Yendred complained to us that he was once again being made to seem a fool. Then, almost casually, Tbạ Hlyak mentioned that the ancient records of "Fishing City" in old Punizla indicate that the Ara Hoot had one more bone segment in each fin than is now the case. This and other evidence had been leading the Punizlan biologists to suspect that Ardean plants and animals were undergoing some kind of long-term changes.

For us it had seemed obvious all along that the life of Arde had evolved. For example, who could miss the similarities between the Ardeans themselves and the Ra Nifids? Not only had life almost surely evolved on Arde, it had probably done so more quickly than life on Earth, other things being equal. The smaller populations within each species would surely have encouraged a rapid spread of advantageous genetic changes.

Are the Ardeans still evolving, or do their brains cancel out further evolutionary progress by ensuring that nearly everyone survives?

Shortly after Yendred left the Lifesystems building, we broke off contact. We told him to expect us at noon of the next Arde-day. This would coincide with our first night session under the new security arrangements.

Thursday, July 10, 9:00 P.M.

In the chapter called "City Below Ground," I mentioned that a steam engine was used in the hadd-rolling mill to drive the rolling truck back and forth over the slab. There had been too little time on that occasion to record it properly or to disentangle its operation within the furious motion of its many parts.

Now, at the Punizlan Institute, we had the opportunity of invisibly meeting its inventor, an old engineer by the name of Tba Grolnp. Near the Machinesystems building was a test pit in which the latest steam engine, a reciprocating two-cylinder model, was undergoing tests. Only about twenty steam engines are currently in use on Arde, but this is not due to their relative unpopularity: the very small size of the Ardean population places a strict limit on the amount of industry necessary to support it. (How many steam engines would there have been in an English industrial town with a population of 40,000 at the turn of the century?)

The one-cylinder model in use at the hadd mill illustrates the heights of Punizlan technical achievement. The reader can hardly fail to appreciate Punizlan mechanical skill if he or she simply recalls that the Ardean mind thinks only in terms of what are essentially one-dimensional scenes. Imagine then, if you can, the internal layout of the steam engine as a succession of invisible chambers receding from your Ardean mind's eye behind a smooth, unbroken outer surface; a three-dimensional engineer could hardly design a better two-dimensional engine!

The engine consists of a chamber in which a piston goes back and forth, pushed west by a spring and east by a burst of steam. The piston operates a driver arm, hinged both on the piston and on the wall of the engine, which wags back and forth with the piston. Steam is generated in a boiler by the tremendous heat of a quite moderate Ardean fire below it. When the valve arm just below the piston slides east, a rush of steam

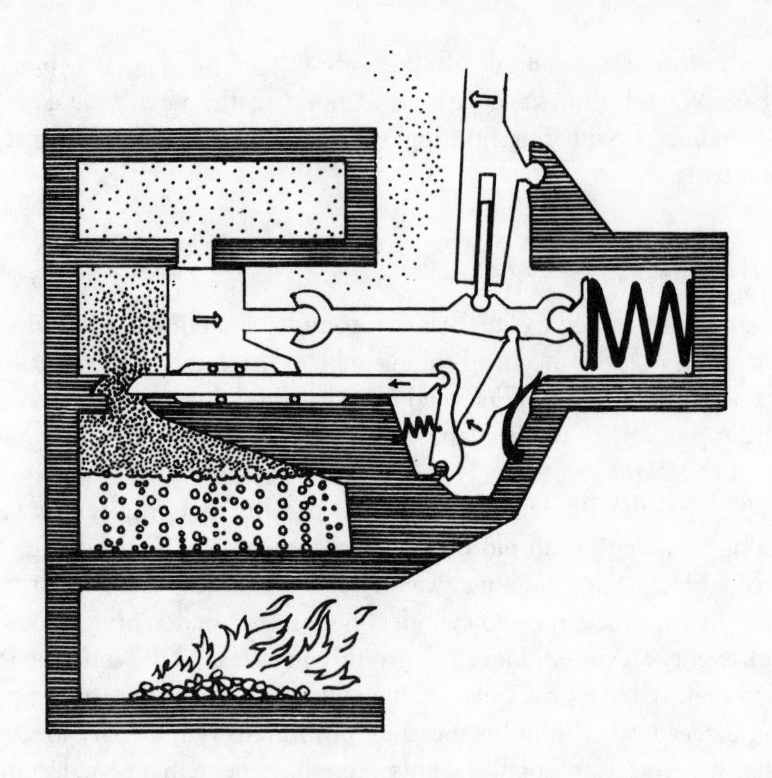

enters the cylinder, slamming the piston eastward against its spring. When the piston passes the entrance to a reservoir above it, much of the steam rushes into the reservoir, immediately causing a drop in cylinder pressure. Even before this, however, the eastward movement of the piston activates a pair of sliding cams, the second of which pushes the valve arm west, closing the valve. Within a split second, the spring has shoved the piston west. It goes far enough to allow all the steam in the reservoir to escape from behind the piston and out of the engine. It also allows the spring on the second cam to slide the valve arm away from the steam valve once more. Again a rush of steam shoves the piston east, and so on.

Compared with our steam engines, this model certainly has some drawbacks. For one thing, it will only run as long as the water lasts. To refill the boiler, the machine must be tilted onto its western side and water poured into the exhaust channel. Operating the drive lever back and forth allows the water to run through the steam reservoir and down through the pressure chamber into the boiler. This operation may have to be repeated several times before the boiler is fully charged.

But for their part, the Punizlans seem totally unaware of shortcomings in their marvelous and useful engine. They have nothing else with which to compare it. It is simply the last word in power technology.

In the same building as that wonderful old engineer Tba Grolnp are two brothers called Rdidn, who, judging from what Ladrd told Yendred before taking him to the brothers' laboratory, were not altogether in their right minds. The first of these brothers had invented a terribly ingenious set of gears for which no use has yet been found. (See the Appendix.) The second brother had been trying for years to invent a wheeled vehicle. He described to Yendred and Ladrd how one day the surface of Arde would be covered with such "motor movers." Yendred objected.

THEY ALL PLANTS AND NSANA WOULD SQUASH FLAT.

NO. ALL NSANA IN MOTOR MOVERS WOULD BE. THE PLANT PROBLEM NO DOUBT SOLVED CAN BE.

HOW THE MOTOR MOVERS EACH OTHER COULD PASS?

THEY PASS NOT WOULD BUT IN THE MORNING EAST MOVE AND IN THE AFTERNOON WEST MOVE.

BUT HOW A DISK TO ROLL BE MADE CAN YET IN THE MOTOR MOVER TO STAY BE MADE?

AH. THAT THE PROBLEM IS.

It must be admitted that the Rdidn brothers' laboratory was a fascinating place. Behind some double wall/doors was a veritable garage full of experimental test vehicles.

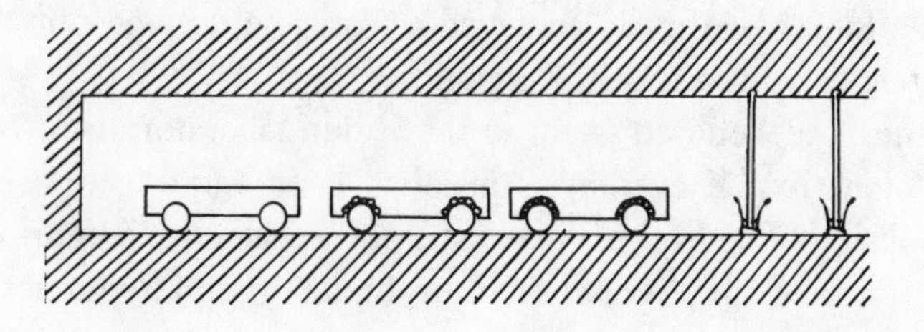

These designs, whose sole object was to get a disk to roll without friction within a fixed location in the vehicle, may all be judged as failures. In the first case the wheel is hopelessly jammed in the hub. In the next vehicle the problem of getting the wheel to roll was "put off" by assuming that the small bearing wheels would roll without friction in their little hubs. But they too would jam. The third design was hardly better. Here the large wheel would turn briefly before the first of the bearing wheels reached the edge of the hub and began to fall out.

The only car built by the Rdidn brothers which could possibly be considered in any way successful was the rather strange vehicle which we found shut away on the next level below this.

This vehicle consisted of a passenger compartment entirely encircled by a tread which rolled on an endless succession of wheels. It was clear that once the passengers had (a) unhooked the tread, (b) entered the compartment, (c) hooked up the tread again, (d) closed the compartment door, and (e) turned on the hrabx, they would have a reasonable hope of a comfortable ride provided that someone could be found to push them the distance. As for getting any sort of motor into such a vehicle, the situation was clearly hopeless.

But, I caution myself, who are we to snigger at the Rdidn brothers? Perhaps at the very moment that I write this, long after our last contact with Arde, they have finally achieved a workable design. Perhaps all real progress begins with a touch of madness!

By the time Yendred's visit to the Rdidn laboratory was over, it was eleven o'clock in the evening. Alice had been wanting to ask Yendred about his visit to Tba Shrin. Finally she got her chance. On his way down to the next level of the Machinesystems building, Yendred described what had happened.

LAST NIGHT I THOUGHT FOR A LONG TIME ABOUT BEING A STUDENT HERE. AT FIRST I WAS VERY HAPPY TO IMAGINE IT. THEN I REMEMBERED THAT IT WAS NOT MY IDEA WHICH MADE TBA SHRIN EXCITED BUT YOURS. WHEN I SLEPT I HAD A STRANGE AND BEAUTIFUL DREAM ABOUT VANIZLA.

▪ WHAT DID YOU SAY TO TBA SHRIN?

SHE ASKED ME MANY QUESTIONS ABOUT MYSELF AND THEN SHE ASKED ME ABOUT VANIZLA. THE MORE I TALKED THE MORE SURE I BECAME ABOUT GOING.

Alice made a V-sign with her fingers. The other students looked pleased.

▪ SO WHAT ARE YOU GOING TO DO?

I THINK I WILL LEAVE SOON. I MUST PAY ATTENTION NOW. EXCUSE ME.

Yendred and Ladrd had entered the computer laboratory and were meeting its inventor. In one end of the room, under a very high ceiling, sat the long-awaited computer, and an impressive device it was. Far too complex to reproduce here (and even harder to explain), it nevertheless consisted of a few simple, tiny components, repeated in seemingly endless arrays throughout the machine. The component itself is easy to display. We even know how it works, thanks to some "directed questioning" by Yendred.

The designer of this incredible machine was called Tba Minh. He stood beside a vertical keyboard and answered his visitor's questions in a lively and energetic fashion. He seemed to jerk, quiver, and bounce as he talked, as though animated by powerful electrical currents. The two problems which he and his assistants had encountered in constructing the computer were (a) supplying power to its components, and (b) getting signals to cross each other.

He had solved the first problem by supplying each component with its own battery, a tremendous feat of miniaturization but one which was quite logical, given the Punizlan dependence on batteries. Their technology had obviously leaped ahead of ours in this regard, for these tiny

THE NAND GATE

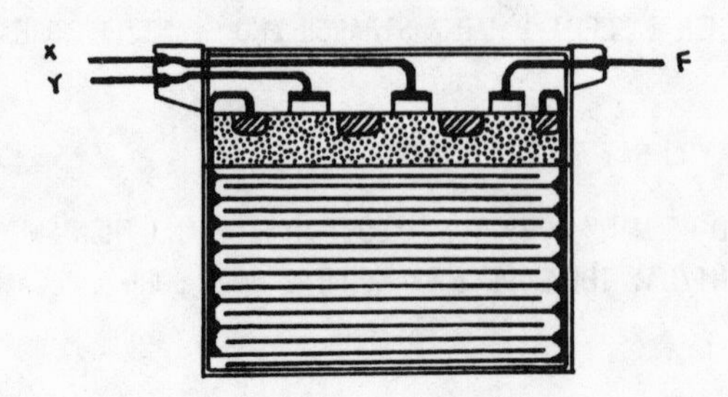

This simple component measures less than a millimeter in width and consists of three parts. The upper part, shown here as hollow, consists of five wires embedded in an insulating material. Two wires enter the device from the left and one leaves it to the right. The two remaining wires are internal. These lead from the central part (which vaguely resembles the cross section of a silicon chip as used in Earth computers) down either side to a battery which occupies the lower portion of the component.

All wires in the Punizlan computer turned out to carry one of two voltages, a moderately small one generated by such batteries and a fractionally tiny one. If we call these voltages HI and LO, respectively, the operation of the device above is very easy to describe.

The two wires on the left, labeled *x* and *y*, carry input voltages. The wire on the right, labeled *F*, carries the output voltage, which is HI unless both the input wires are HI. In the latter case, the output voltage drops to LO. This logical operation turns out to be exactly what Earth scientists call the NAND operation: the output is HI if "Not both *x* AND *y*" are HI.

From this one device it is possible to construct logic circuits, registers, loaders, clocks, memory, and all other parts of a digital electronic computer. The Punizlan computer, however, appeared to have no clock circuits, its various components all operating asynchronously.

batteries would last for several years. The other problem, that of getting the signals to cross, had been solved by mimicking the Ardean brain, using the electronic analogue of the biological crossover circuit. (See the Appendix.)

Having continually described his computer as an artificial brain, Tba

Minh at last invited his young visitors to submit a sample problem for it to solve. He bounced up and down with anticipation. Then Ladrd described a problem.

> THREE NSANA THREE BAGS ARE CARRYING. THE FIRST BAG THREE MORE BREADS THAN BASLA HAS AND THE SECOND BAG [. . .]
>
> I THAT KIND OF PROBLEM DID NOT MEAN. THE MACHINE ONLY ARITHMETIC DOES.
>
> I APOLOGIZE. THEN EIGHT TO THE EIGHT TO THE EIGHT CALCULATE.

The good scientist flapped his arms with embarrassment and remarked that the machine was not able to contain the enormous number which would result from this calculation. Yendred interrupted.

> YOU THIS LIKE A BRAIN SAY IS? HOW THE BRAIN OF LADRD EIGHT TO THE EIGHT TO THE EIGHT UNDERSTANDS BUT YOUR MACHINE DOES NOT?

Tba Minh, bouncing and quivering a good deal less, explained that such machines were at an early stage of development but that sooner or later they would be able to outthink the Ardeans themselves. While Ladrd took up the thread of this conversation, Yendred spoke privately with us.

> IS YOUR COMPUTER LIKE THIS MACHINE?
>
> ▪ YES, BUT MUCH MORE ADVANCED. TBA MINH AND SCIENTISTS LIKE HIM HAVE MANY PROBLEMS TO SOLVE BEFORE YOUR COMPUTERS BECOME LIKE OURS.
>
> IF WE HAD AN ADVANCED COMPUTER COULD WE CONTACT YOUR WORLD?
>
> ▪ I THINK THAT WOULD BE MOST UNLIKELY.
>
> ARE YOUR COMPUTERS LIKE BRAINS?
>
> ▪ NOT REALLY, BUT EVERY YEAR WE FIND NEW THINGS FOR THEM TO DO. HOWEVER, WE ARE MORE IN DANGER OF HUMANS BECOMING LIKE COMPUTERS THAN COMPUTERS BECOMING LIKE HUMANS.

Tba Minh finally got a workable problem from Ladrd and fed two four-digit numbers to the machine on a vertical keyboard. The answer came back in about three seconds on a scroll of tape printed upon by a most ingenious two-dimensional typewriter located just below the keyboard.

Yendred and his friend visited the Ardesystems building next. Here it was that we probed two scientists about their knowledge of their planet's interior. They had long ago deduced that Arde's ocean crust was slowly moving toward the continent and had elaborated a well-developed theory of Ardean tectonics as a result. (See the Appendix.) Yendred finally asked us to cut our questions short as there was yet another building on their day's tour. This building housed Largesystems, the Punizlan equivalent of astronomy and cosmology.

Fully half the personnel of Largesystems were away on Dahl Radam at the institute's observatory. Among those who remained was a young female astronomer named Kewbt, who showed Yendred a collection of ancient Ardean astronomical instruments. As she went from sighting pole to telescope she reviewed the history of astronomy in Punizla. Until this time the seasons of Arde had mystified us; how could Arde, with no hemispheres, have seasons? Kewbt demonstrated for the visitors Arde's orbit around Shems by picking up two disks and, hooking the larger to the ceiling, simulating one-half of Arde's orbit by swooping across the room, holding the smaller disk aloft. This disk ended up twice as far from Shems as it had started out. If this were a true representation of the eccentricity of Arde's orbit, it would easily explain the difference between summer and winter.

We persuaded Yendred casually to inquire into the exact amount of eccentricity, and Kewbt cheerfully informed him that it was approximately .52. Long after the contacts ended, we used this figure to plot a reasonably exact orbit for Arde. The orbit, in turn, enabled us to calculate that in a little over seven sidereal years, Arde experienced exactly ten seasonal years. In other words, after going around Shems slightly more than seven times in relation to the background of fixed stars, Arde approached and then receded from Shems a full ten times. This situation contrasts markedly with Earth, where the two kinds of year are practically identical. How different from Earth's simple ellipse is the wild, weaving orbit of Arde! The figure below is taken from a plot of Arde's orbit made by two colleagues of mine at IBM.

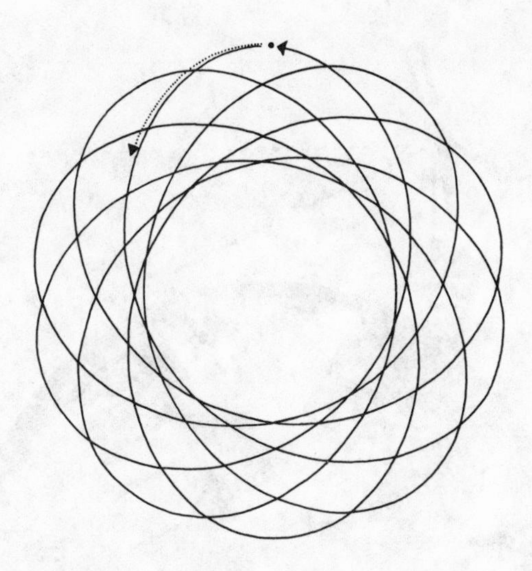

This strange, braid-shaped orbit is due to the different force of gravity in the Planiverse: as explained earlier, it does not lose its strength so rapidly with distance as the gravity in our universe does. From the diagram it might appear that Arde is about to retrace a pattern of ten seasonal years and seven sidereal ones. But not quite. In the midwinter of every tenth year, Ardeans—both Punizlans and Vanizlans—celebrate the "day of harmony." On midnight of that day, the sky looks almost exactly as it did ten years previously; the seasonal and sidereal year are once more in harmony and the constellation called "Meezn" is directly overhead. It is counted as especially fortunate if this constellation can actually be seen through the clouds of a midwinter night. Unfortunately, as decade follows decade, Meezn is not in quite the same place as it was before. This creates endless bother for Punizlan astronomers, who, every thirty years or so, must insert an additional year in the ten-year cycle. A traditional Punizlan preoccupation with determining the ratio of the two years has culminated in the currently accepted value of .698128.

The Largesystems building also turned out to be the headquarters of the Punizlan space program. We knew that the Punizlans had long ago developed rocket aircraft, but here we first learned that they had put satellites in orbit and were launching a shuttle rocket to an orbiting space station every two weeks! On the third floor of the building was a model of the station.

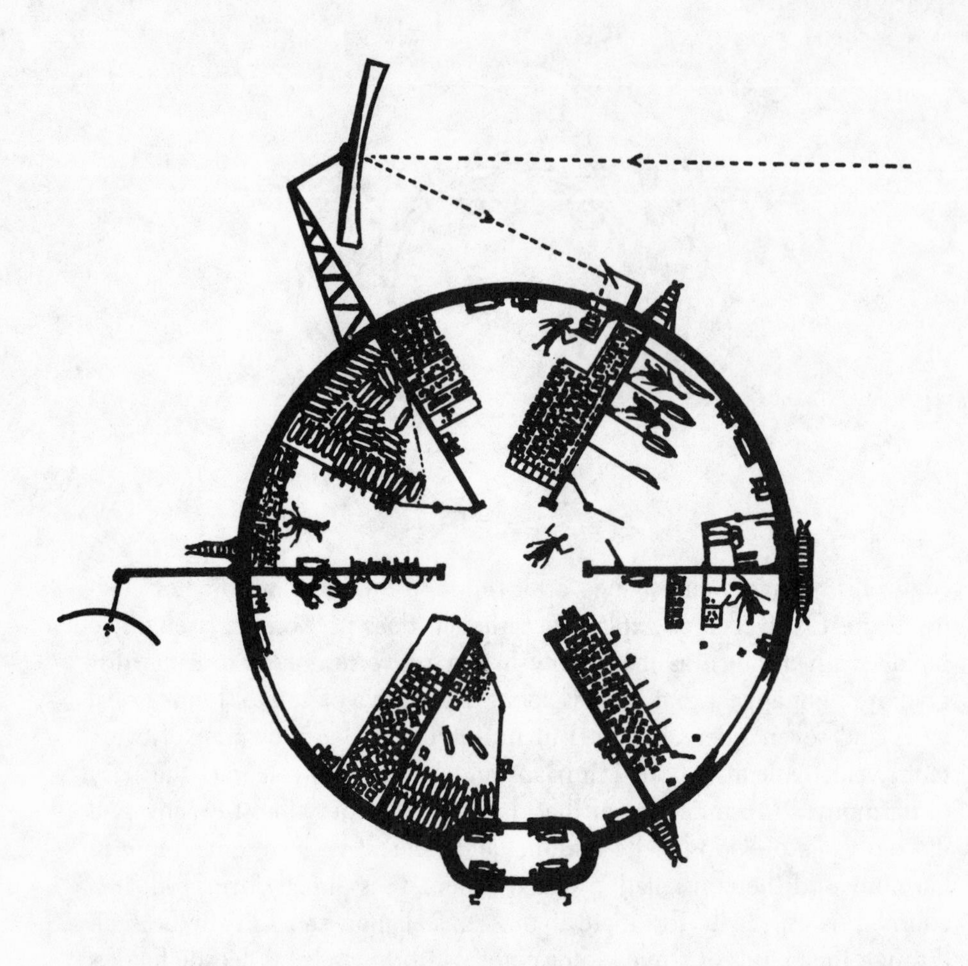

From a launch facility not far from Maj Nunblt, a series of six "module rockets" had been launched into orbit, each having a wedge-shaped portion of the station as its upper stage. These were assembled in space into the station itself.

Kewbt took the model of the station apart and explained its various parts in detail. She had two rapt listeners that she was aware of and six more besides.

Every few hours the Punizlan space station circles Arde. A crew of six scientist-astronauts glide easily throughout the interior of the station, experiencing a freedom of motion which almost hints at the experience of three dimensions. From any of its three windows the occupants may look out into the cold immensity of Planiversal interstellar space. Perhaps as he examined the model before him, Yendred imagined himself to be floating, weightless behind a cold, clear jebb window and staring out at that vast circle of coldly flaring stars, wondering in what direction the Earth was. He would know that it was in no direction he could see, but he would look and wonder anyway. Below him he would at least see his own planet, Arde; a strip of brightness veiled by wispy clouds.

The space station has its own atmosphere, supplied from pressurized bottles mounted around the inner wall. Electric heaters keep the interior warm, and the station is well stocked with a twenty-four-day supply of air bottles, food, batteries, and other necessities. Every sixteen days a shuttle rocket arrives, bringing a new astronaut and new supplies. The shuttle docks at the station by means of two latches on either side of the air lock. Luckily, it is absurdly easy to make a perfect seal in the Planiverse—as when the Punizlan farmer inadvertently created one in attempting to lift the barn cover at one end.

After a 107-minute social hour, the shuttle undocks and begins its gentle ascent to Arde, taking with it used batteries, spent air bottles, and the Punizlan astronaut who has been on board the station the longest.

Three major pieces of equipment on the station are a giant space telescope, a simple on-board computer, and a radio for communication with Arde. The telescope has a six-meter almost-parabolic primary mirror fixed to a strut. Parallel rays of light striking this mirror at a slightly off-axis angle are reflected to a secondary mirror and thence through an (optical quality) window and into a camera. Thrusters outside the station can change the station's orientation so that the telescope may examine whatever remote reach of the Planiverse its operators wish to view.

Yendred had to pass up the opportunity of visiting the launch facility just twelve kilometers east of Maj Nunblt. Upon emerging from the Largesystems building, he and Ladrd saw Tba Shrin approaching hastily. In her upper eastern arm she held what appeared to be a small strip of paper.

THIS A TICKET FOR YOU IS. I OF IT HAVE NO NEED. A ROCKET FOR SEMA RHUBLT SOON LEAVES.

Yendred was quite touched by this gesture and, as was common for him in emotional situations, he overreacted.

I THIS DO NOT DESERVE. I NEITHER INTELLIGENT NOR WORTHY AM. ALL CREDIT ELSEWHERE LIES.

YOU PERHAPS ONE DAY TO MAJ NUNBLT RETURN WILL. WHAT YOU IN VANIZLA SEEK I YOU WILL FIND HOPE.

It was about 3:00 A.M. in our laboratory as Yendred carried his few possessions in a string bag from Ladrd's apartment to the rocketport just outside Maj Nunblt. We decided to keep the contact going as far as possible into our own morning in order to witness the rocket trip. Alice, at the terminal, seemed just as excited about it as Yendred.

I HAVE NEVER RIDDEN IN A ROCKET PLANE BEFORE. THEY ARE EXTREMELY FAST BUT SAFE. THEY ALMOST NEVER CRASH.

▪ I WISH I COULD RIDE WITH YOU.

BUT YOU CAN SEE EVERYTHING THAT GOES ON.

▪ IT IS NOT THE SAME THING AS BEING THERE.

8

Traveling on the Wind

The rocketport was a half-hour walk to the east of Maj Nunblt. When Yendred arrived, the wind was blowing hard from the west and bits of material, refuse from the city, flitted along the ground, leaped over his person, and swirled briefly in his wake. He stopped at the entrance to the ticket office to inquire about the next flight and discovered that there would be a delay due to a stuck fuel cartridge. Most of the other passengers had already boarded and he was advised to do the same.

We had wondered how an aircraft of any kind could fly in Arde's atmosphere. After all, wings would be completely useless, merely acting as barriers to the slipstream. When we saw the craft that Yendred was about to board, it became clear that the whole rocket plane was itself a wing.

Yendred mounted a walkway into a hatch in the front of the rocket plane and took the furthest available seat, this being a social rule for any public facility on Arde. The interior of the plane had two passenger compartments. The upper one had its full complement of four passengers and the stair to it had already been swung upward and locked in position.

The pilot sat in a sort of elevated cockpit, but how he would control the flight of the craft was not at all clear. In the rear of the rocket plane was an enormous, articulated triangular section with a hollow channel running down its middle. Workers appeared to be pushing on a large casing filled with some powdery solid, trying to force it into the channel. This was the fuel cartridge, filled with inflammable powder the force of whose ignition would drive the rocket into the sky. But how would it be steered? Certainly, there would be no problem about lateral motion since the Planiverse has no width. But how would the pilot control the rocket's flight angle once it was airborne?

While waiting to see how this would be accomplished, we examined the rest of the craft in detail, noting a luggage compartment at the rear and noting also that each seat on the floor of either compartment was matched by a corresponding seat on the ceiling.

Ffennell leaned close to the screen, studying the rocket plane intently. "Do you think that thing flies upside down, too?"

The rocket plane was pointed west, opposite to the direction of Yendred's destination, so it seemed to us that some inverted flying would be necessary. With a few blows of a hammer, one of the workers succeeded in getting the fuel cartridge past its stuck position and the crew then easily lodged it in place. A last passenger boarded and a member of the ground crew picked up the entrance ramp and carried it inside the hatchway. He passed one end of it up to the pilot, who inserted its lower end in a socket just forward of the hatchway. This seemed a rather odd thing to do with an entrance ramp. Why didn't they simply store it on the floor of the craft? Our question was almost immediately answered as the pilot descended from the cockpit and took the end of a cable handed to him by one of the ground crew. Clipping it to a point not far from the base of the ramp, the pilot ascended to his seat and took firm hold of the ramp. It was a control stick! The cable ran all the way to the engine housing at the rear. Now everything was clear: when the pilot pushed forward on the stick the housing would be tilted downward, directing the engine's thrust lower, causing the tail of the craft to rise and the rocket plane itself to dive. Pulling back on the stick allowed a spring (attaching the top of the engine housing to the fuselage) to pull the housing upward, with the opposite effect.

As if by some signal, all the passengers reached toward the ceiling and pulled down elastic cords, which they secured to the floor behind their seats. By now the rest of the ground crew had disappeared—except for one hardy soul whose duty it was to light a two-meter fuse which hung from the opening of the fuel cartridge. He did so, then ran with a sort of panicked wobble toward an entranceway not far off.

The rocket nudged forward slightly and suddenly it just wasn't there any more! Lambert gaped at the screen.

"It blew up," said Edwards.

"No," said Chan. "It's just fast as hell. Lambert, set the scan speed at 1000 and back the scale off about twenty. West and hurry!" Already the rocketport scene had begun to wobble, sign of a weakening Earth/Arde link.

The rocket plane took a seeming eternity of frantic key-punching to find and then to track. One moment we had a glimpse of it zooming west and then we found it a few seconds later traveling straight up. It was obviously "turning" east. We finally discovered the right speed and direction, locking onto the craft with our screen. It showed up as a tiny teardrop streaking eastward at an altitude of roughly 12,000 meters. Cautiously we zeroed in until once more it loomed in our screen.

▪ YENDRED. HOW IS IT?

AT FIRST I THOUGHT THE MOTOR HAD EXPLODED AND KILLED US ALL.
NOW THERE IS A HORRIBLE SCREAMING SOUND.

▪ THAT MUST BE THE MOTOR. HOW IS YOUR FLIGHT?

ARE WE FLYING?

What amazed me was how the passengers could have withstood such acceleration. Did the Ardean exoskeleton act as a kind of pressure suit?

The flight lasted about twenty minutes, Lambert becoming quite adept at following the occasional rises and falls in the rocket plane's position. He had learned to keep his eyes on the pilot, so that when he began to pull back or push forward on the stick, Lambert would have a split-second warning about which way the rocket plane would go. The air at this altitude was relatively calm, lying as it did between the lower, westward-moving layers and the upper, eastward-moving ones. I assumed that the amount of fuel in the rocket engine was carefully measured so as to run out during the landing in Sema Rhublt.

Without warning, the pilot pushed sharply forward on the stick and the craft dropped off the screen. When Lambert located it a few moments later, it was heading west. During the eastward flight the passengers had all climbed into the ceiling seats across their restraining cords and now were barely settled, once again, on the floor seats. During the flight, the cords had all been undone of course, but now they were stretched taut once again in preparation for landing.

The pilot had evidently timed the rocket plane's descent so that it would run out of fuel at the exact moment of landing. Having completed his westward turn, he leveled the craft, even as the ground rushed up to meet it. A few underground houses and a lone Ardean on the surface flashed by.

From what we could see of the fuel cartridge, there was no more propellent left, yet still the craft streaked along, just above the ground. Abruptly it began to drop more quickly and, just before it would have jarred into the ground (possibly doing real damage), the pilot pulled back hard on the stick so that the engine housing now acted like a huge, passive aileron, bringing the nose up so that the craft glided in for a beautiful one-point landing.

Such is the wonderful economy of Punizlan technology that not only the control stick but the control cable had two uses. Here, during the landing, it acted as a skid, protecting the skin of the fuselage from any damage.

I took over at the keyboard.

▪ DID YOU ENJOY THE FLIGHT?

IT SEEMS UNREAL. I FEEL LIKE EVERYTHING IS A DREAM.

The passengers unhooked their cords and stood up over their seats. The pilot unhooked the control cable, and a member of the ground crew took the control stick from the pilot and converted it back into a ramp. The passengers then began to disembark. Yendred was the second passenger to emerge into the bright daylight of a new city.

THIS IS SEMA RHUBLT. MANY ARTISTS AND ARTISANS LIVE HERE.

▪ WE MUST LEAVE YOU NOW IN ORDER TO SLEEP. WE WILL BE BACK EARLY IN THE MORNING TOMORROW.

It was 4:00 A.M. when we killed the 2DWORLD program. It had been a very long session and there were a good many notes to collect and photographs to file. The rented video camera had developed a flaw. I would have to replace it before coming to work that afternoon. We carried everything to my office and then left the building by different routes.

Friday, July 11, 9:00 P.M.

The next evening we were back in front of the terminal, watching Yendred eating breakfast with a family of Punizlans in Sema Rhublt. They had taken him in as a guest and persuaded him to stay a few days before traveling on to Dahl Radam, the great central plateau whose gradual rise began only three hundred kilometers to the east. In spite of the attractions of Sema Rhublt, its industries, artists and artisans, it was uncharacteristic of Yendred to agree to such a long stay. In any case, we had learned to be patient and to let the events of Arde unfold as they would.

Besides two parents, Yendred's host family consisted of five children ranging from a daughter of roughly Yendred's age down to a son barely five years out of the egg. Yendred sat between the father and daughter as they ate.

As a rule, Ardeans take their meals in silence. We think that the reason for this is that the stomach gradually fills up with food, cutting down on the supply of air available for speech by the jaw zipper muscle. Yendred used the silence to tell us he was getting on quite well with this family, who were distant relations on his mother's side. Great plans were in store for a tour of Sema Rhublt, including a visit to Dar Jisbo, one of Punizla's greatest artists.

Presently, the father walked over to the stove to allow Yendred access to the swing stair. In turn, each of the children asked, and received, permission to accompany Yendred on the tour.

They made an amusing sight as they climbed the entrance stair, Yendred first and the smallest child last (his name was Trab), barely able to take the steps with his tiny legs. Once on the surface, they seemed a curious troupe. An Ardean approaching them would see only Yendred, while an Ardean following them would see an almost perfectly graded sequence of heads.

Soon they descended by another stairway to a sort of crafts bazaar, where a great many artisans were busy making bowls, utensils, shelves, figurines, rope, and small amusing machines. The eldest daughter, Na by name, carried with her a string bag in the bottom of which was some Punizlan currency. Yendred stopped to watch a bowlmaker at his trade.

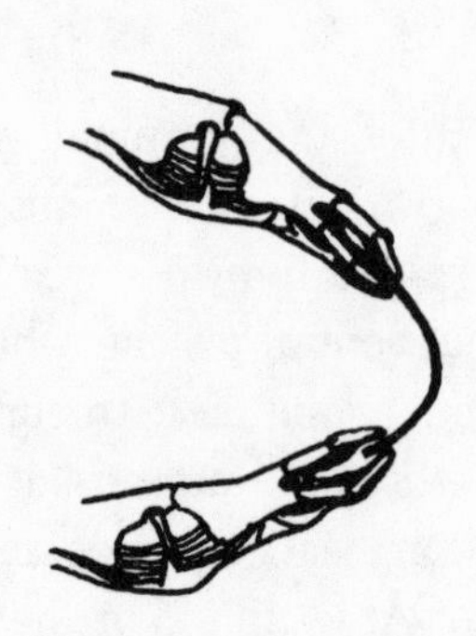

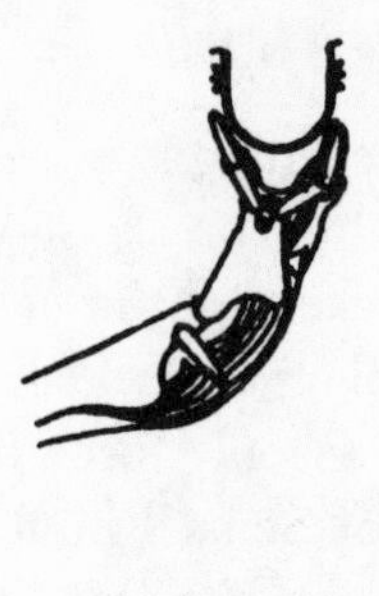

Taking what appeared to be a length of wire, the artisan first bent it into a U-shape and then filled it with water, studying it for a time and then emptying the water. Making a few adjustments in the shape of the cup, he next attached a base, by swabbing the bottom of the cup with glue and pressing it firmly into the base. Next he glued on some ornamental bosses and then carefully painted the cup, inside and out, occasionally changing brushes. Finally, he set it into a small oven in the bottom of which were just a few coals, these being sufficient to bring the cup to a very high temperature in Ardean terms.

THAT BROWN CUP HOW MUCH IS?

HERE OTHERS JUST LIKE IT ARE. EACH ONE TWENTY-THREE JEB IS. HOW MANY YOU LIKE WOULD?

PLEASE ME SEVEN GIVE. TWO LARGE AND FIVE SMALL.

Yendred reached into his glued-on pocket and retrieved some jeb coins. Finding these not enough, he also took out a bill.

▪ YNDRD. IT'S NONE OF MY BUSINESS, BUT ARE YOU GOING TO HAVE ENOUGH MONEY LEFT FOR YOUR TRIP?

OH YES, MORE THAN ENOUGH. IS THIS DEWDNEY?

▪ YES. I HOPE YOU DON'T THINK I'M INTERFERING.

IT IS MOST KIND THAT YOU ARE CONCERNED. HOW MUCH IS IN MY POCKET?

▪ I SEE TWELVE BILLS AND FIFTEEN COINS.

THAT WILL CERTAINLY BE ENOUGH FOR A FREIGHT BALLOON.

I then asked Yendred a question which I had been putting off for some time. How did he intend to travel home from Dahl Radam?

THE GOING IS ENOUGH. I AM NOT CONCERNED ABOUT RETURNING. WHY SHOULD YOU BE CONCERNED?

Yendred and his entourage left the bowlmaker's booth, little Trab leading the way. After a hurried consultation with Na, Yendred issued the orders about which way to go—up or down. Each of the children whispered to the one ahead until the message finally came to Trab. Down.

Down they went to the next booth. This one belonged to an ancient hboolmaker, hbool being the Ardean equivalent of rope. Now everyone had a chance to watch as the old Ardean deftly glued specially treated fibers of Basla plants together. He started with a row of single fibers all laid end to end.

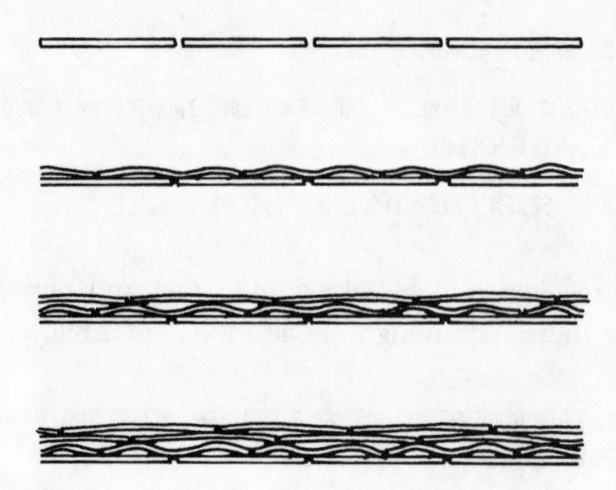

He then glued down another sequence of fibers upon these, using a staggered pattern like house bricks on Earth. At each stage he was careful to leave a small air pocket between adjacent fibers, since this is what gives hbool its ropelike flexibility.

Hboolmakers also turned out a product somewhat like plywood. Using the same plant fibers, they simply run a strip of glue along the entire length of the "rope" before laying down the next row of fibers. After a few days, when the glue has hardened, the finished product is much more like a piece of wood than a rope. It may be used to make shelving, brackets, supports, and all manner of furniture.

According to Yendred, making hbool, bowls, and other objects was almost a lost art in Punizla, as all these items were now mass-produced in factories. In fact, as Yendred learned, the hboolmaker they now watched had been "imported" from Vanizla.

YOU HOW LONG IN SEMA RHUBLT HAVE BEEN?

TWENTY-TWO YEARS.

PARDON?

TWENTY-TWO YEARS AMONG THOSE WHO HEAR CANNOT.

We were to learn later that although Punizlans and Vanizlans speak the same language, their pronunication and vocabulary differ somewhat.

LAYERED MACHINES

The plywood-like material constructed layer by layer is called "hatr" by the Ardeans. An especially strong form of hatr is made from thin strips of hadd. For working hatr and joining pieces of it together, the Ardeans have an extremely limited tool kit: saws are impossible, of course, and the old familiar hammer-and-nail combination is out of the question.

To give a piece of hatr a specific shape, about the only tools available are the hammer and chisel. More often than not, however—at least in Punizla—complicated shapes are produced by laying down the strips directly in the required pattern.

This technique becomes a bit tricky when constructing machinery, however. Consider the example of the disk-and-socket hinge.

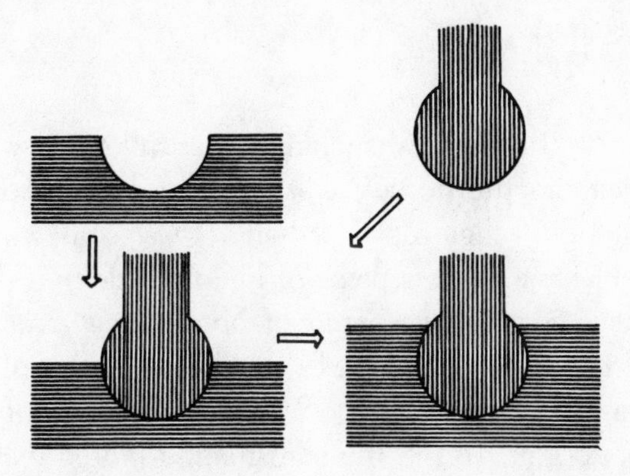

Here, the socket is at first only partially built up. Once the disk is inserted, the rest of the socket may be completed. For most machines it is not a question of assembling hadd-hatr components but constructing the components along with the machine.

Another building material in frequent use on Arde is jebb, a glassy rock which is quarried on Dahl Radam. It may be shaped by chisel or melted and poured into molds. A form of jebb-hatr is used for windows.

On this occasion, Yendred was probably having trouble with the old Vanizlan's dialect.

YOU HBOOL HAVE MADE ALL THIS TIME?

YES. TO YOU PUNIZLANS IT A CURIOSITY IS BUT TO SOME A LIVELIHOOD IT IS AND A NOBLE ART.

IT A CRAFT IS, IT NOT IS?

IT A CRAFT IS ONLY TO THOSE WHO SEE CANNOT AND WHO FEEL CANNOT.

WHERE THEN THE ART IS? IF I STUPID SEEM, ME PLEASE FORGIVE.

VERY WELL. I HERE TWO PIECES OF HBOOL MAKE.

I ONLY ONE SEE.

AH, BUT THERE TWO ARE. ONE OUTSIDE IS AND THE OTHER INSIDE IS.

THE HBOOL THE HBOOL INSIDE IS?

NO. THE OTHER INSIDE ME IS.

I had been waiting for just such a time as this when we would finally meet a Vanizlan "in the flesh," as it were, and learn something about their philosophy and religion at first hand. I was somewhat mystified by the old chap's claim to have a piece of hbool inside him. Looking at our screen, we knew he hadn't a shred of hbool anywhere in his innards. Recalling Tba Kryd's description of the Vanizlan habit of thinking symbolically, I then wondered if the hbool being worked on somehow symbolized some process going on in the craftsman. Yendred may have had the same problem, because he silently watched the hboolmaker for a while.

Soon enough, however, he began another conversation, this time about art. The old one seemed to have warmed slightly to Yendred's interest and launched into a curious, rambling discussion of the different forms of art in Vanizla. There is frequent use of repeated patterns—friezes, one might call them—consisting of the same image copied serially for a great length. The other sort of art involves a pattern which almost repeats, changing subtly from one image to the next. This is intended to represent either some sort of spiritual transformation or the representation of a higher form of reality: the old Vanizlan became rather vague on this point.

The other major theme pursued by Yendred in his conversation with the hboolmaker concerned those with "knowledge beyond thought." It was here that Yendred's search for the Vanizlan called Drabk received some direction.

YOU OF DRABK KNOW?

YES. DRABK THE SHARAK.

WHERE I HIM MAY FIND?

SOME BY THE SHRINE OF AMADA SAY AND OTHERS IN THE TOWN OF OKBRA SAY. WHY YOU HIM DO SEEK?

I THE KNOWLEDGE BEYOND THOUGHT TO LEARN WISH.

PERHAPS THE KNOWLEDGE BEYOND THOUGHT BEYOND YOU IS.

During this conversation, Na's siblings had grown increasingly restive, shuffling their little feet from side to side and engaging in small nudges and whispered remarks. Na enclosed Yendred's seven cups in her string bag, handed it to the eldest child, and sent the whole troupe home. Their next visit could hardly entertain the children any more than the hboolmaker had.

Yendred and Na then traveled on together to the home of Dar Jisbo, probably Punizla's greatest artist. When the two arrived at his house, the great artist was working at a new painting. He asked his young visitors to make themselves comfortable. They were free to watch as he finished a critical portion of the work at hand.

It was quite absorbing to watch a one-dimensional picture being painted. On the easel was a board and on the board was a thin strip of material taped at both ends onto the board. Dar Jisbo would hold the brush above the surface, make a tentative dab, and then lower it again to inspect the image taking shape. Of course, he could only see over his upper arm, the one that held the brush. His lower arm held a small tray of paints. The extreme delicacy of his movements and the judicious air with which he regarded his work made us long to see the picture itself. But, in this as in so many other cases, the layer of pigment was too fine to resolve without taking the time to scan the image at high magnification.

Yendred and Na had been inspecting a collection of sculptures at one end of Dar Jisbo's studio. These were mostly small and were set on shelves. Whether they were made by Dar Jisbo or someone else, we never discovered.

From the sculptures, the young visitors turned their attention to the ceiling of the artist's studio, examining it with great interest.

HERE TO THE CEILING GLUED A PRINT OF THE MOST FAMOUS PAINTING IS.

■ ARE YOU TALKING TO US OR NA?

TO YOU. EXCUSE ME PLEASE. HERE ABOVE US IS A FAMOUS PAINTING REPRODUCED.

Yendred and Na were looking up at a long, thin strip of paper glued to the ceiling. It seemed to me that the time had come to scan one work of art, at least for the sake of the record. Even as we talked to Yendred, we painstakingly scanned along the strip of pigment that formed the print's surface. We recorded the subtle variations in thickness as we went, hurrying to finish when Dar Jisbo announced that he was at last ready to talk.

The result of our scan is produced below. In order to save space, I have turned the picture on its side; moreover, I have widened it for easy visibility to our own, three-dimensional eyes.

The painting's title translates roughly as "Lady with Basket of Kobor Hoot," a work of the great Fls Bwit, a Punizlan artist of the previous century. I have already mentioned how difficult it is for us three-dimensional beings to appreciate scenes from an Ardean point of view. (One might suppose that four-dimensional beings would find the scenes that *we* view equally incomprehensible.) After long analysis, the students and I located a middle section of the picture which stands a fair chance of being the basket of Kobor Hoot and another section to the right of this which is probably the lady's head. But even if we had recorded the picture with complete accuracy, to us the whole thing would still resemble a spectograph of Arcturus.

In spite of such difficulties, I have speculated a little on the problem of representing two-dimensional objects on a one-dimensional surface and have arrived at a rather startling conclusion about Ardean art: each representation of an object by an Ardean artist *automatically* looks like a great many other things besides the object being represented. Moreover, distance relationships are far less clear and the whole form is already rich with the suggestion of many other things, as though a human painter could form an image which looked simultaneously like a barn, a vase, and a kangaroo. Ardean art is inherently surreal!

It took very little prompting on Na's part to get Dar Jisbo to speak about his work in particular and art in general.

I TO THE ESSENCE OF MEANING OF FORM HOPE TO GET. I FROM THE BONDAGE OF PERCEPTION AND FROM MY MENTAL PRISON SEEK TO ESCAPE BY NOTHING PAINTING.

HOW NOTHING DO YOU PAINT?

I RANDOM FORMS CREATE IN THE HOPE OF SOMETHING WHICH LIKE NOTHING ELSE LOOKS OBTAINING. BUT ALWAYS I A SUGGESTIVE FORM OBTAIN. IT MOST FRUSTRATING IS.

BUT OF YOUR FAMOUS PAINTINGS WHAT? OF YOUR ''FORMLESS FORM NUMBER SEVEN'' WHAT?

FAILURES, ALL FAILURES. FORMLESS FORM NUMBER SEVEN ME PERSONALLY REMINDS OF A RA NIFID WITH A STEAM ENGINE LOVE MAKING. ALL FAILURES.

PERSPECTIVE AND AMBIGUITY

The simplest, precise demonstration that objects tend to look like each other to the Ardean eye comes from considering a square and a rectangle. There is a point of view from which both will appear to be the same object.

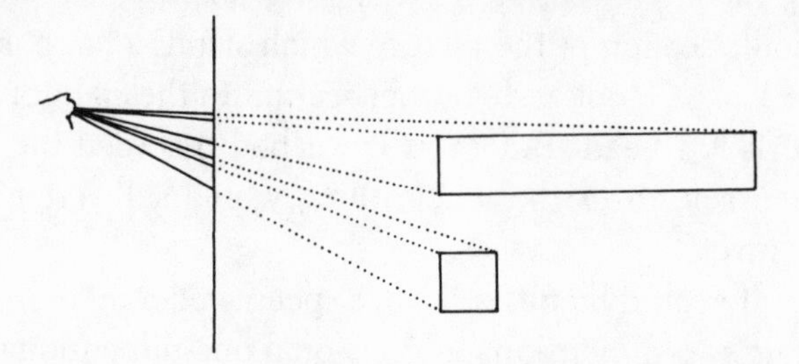

If we imagine an Ardean artist painting the geometric still life above, an accurate placement of these images on paper would correspond to the intersections of lines from the objects to the eye with a vertical line representing the paper. It is easily arranged (and accidentally inevitable) that the object drawn on the paper to represent the square looks exactly like the object drawn to represent the rectangle. This is just another way of saying that in this case the eye cannot tell these objects apart.

A similar, three-dimensional demonstration involving a cube and a long box would appear to be extremely difficult (if not impossible) to arrange.

Na did her very best to shake Dar Jisbo from his despondent mood, ending by suggesting that he paint Yendred's portrait.

> YOU, DEAR ONE, CLOSER ARE AND FAR MORE BEAUTIFUL. ALLOW ME WHAT LITTLE OF MY CONVENTIONAL SKILL REMAINS TO DEMONSTRATE.

Taking a small pad of paper, Dar Jisbo studied his subject carefully and then, beginning at the top of the paper, skipped his pen lightly and confidently along its surface from top to bottom. The whole operation took about ten seconds. At the end, he detached the paper and handed it to Na, who cocked her head to examine it. Yendred leaned eagerly over her, the better to view the portrait.

IT PERFECT IS! WHAT SKILL! WHY YOU MORE PICTURES LIKE THIS DO NOT CREATE?

IT TOO EASY IS.

The great artist sounded almost sad as our printer clattered out this last sentence. We watched as Yendred and Na bade Dar Jisbo farewell.

It was now our turn to say goodbye to Yendred, agreeing to "meet" him after sunset on the next Arde-day.

Sunday, July 13, 9:15 P.M.

This was to be Yendred's second to last evening in Sema Rhublt. He had earlier purchased a ticket for a freight balloon which would depart in two days. He and Na were part of a procession of Punizlans walking to the west in Sema Rhublt. Between snatches of conversation with Na, Yendred filled us in on his day.

They had been to something like a museum, visited a "flinka farm" (whatever that was), and had ended up at the Punizlan equivalent of a restaurant before returning to Na's home. Shems had just set and they were on their way to a concert.

The concert was held in a subterranean amphitheater whose covering cable had been removed, opening all the seats to the evening sky. The audience, including Yendred and Na, sat down in the order of its coming. Each new arrival stepped over seats to get to his or her place and then sat expectantly as the orchestra tuned up its instruments.

The orchestra was very small by our standards, consisting of just seven musicians. Presumably, they would produce between them enough sound to reach even the highest seats, those now being filled by the latest arrivals. Presently the amphitheater was full and attendants at either rim mounted an odd-looking battery-operated lamp which beamed down on the stage. I have often recalled that scene, imagining it all from an Ardean point of view: the freshly waxed and polished bones of the musicians would gleam in the artificial light. The strange sounds of even stranger instruments would drift up over the squeaking babble of the audience. Unearthly bonging, twanging, and humming notes would sound their briefly rising glissandos and trail off into the newly emerging stars overhead. In this one scene I find more coziness than confinement, perhaps because Yendred was there.

The instruments of the orchestra, when we examined them closely and watched the performers tune them, were difficult to classify by our own standards: was a string stretched between two posts a drum or a violin? The only beings untroubled by such questions were the Punizlans who now awaited the beginning of the concert.

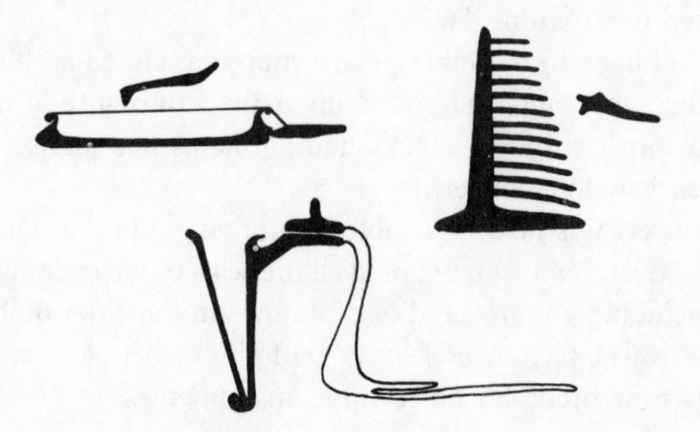

Yendred kindly took some time off from his conversation with Na to explain briefly the names of three of the instruments and how they worked.

The bwiki consists of an open frame with a wire strung tightly over it. Firmly attached to the frame at one end, the wire ends in a movable lever manipulated by the bwiki player with one hand. In her other hand, the musician holds a small hammer. A mere peck on the wire produces a tone whose pitch depends on the pressure exerted on the lever. If the hammer is held on the wire after the blow, however, its position on the wire now becomes critical as a pair of tones is produced, hopefully in harmony with each other. Striking the wire with rapid, fluttery pecks and firm, jerking strokes, the bwiki player creates intricate, tiny melodies interspersed with powerful, wailing chords.

There were two bwiki players. Lambert remarked that they must be "first and second bwiki." At least he was enjoying himself.

The musmar is rather like the "thumb piano" used by certain cultures here on Earth. A graded sequence of metallic rods is fixed in a vertical stand, and each rod, when struck by a special hammer, produces a fixed note in the basic Ardean scale. It probably goes "bong."

The ma eeryeh is the third instrument in the ensemble, consisting of a V-shaped apparatus partially filled with water. From the eastern side of the V protrudes a ledge attached to a long string, at the other end of which is a rod with a handle on it. The ma eeryeh is essentially a bagpipe: the musician deftly gathers air into the string bag and claps the rod onto the ledge, trapping the air inside the bag, which is held between the two eastern arms. A gentle pressure on the bag creates a flow of air which whistles as it passes the ledge. The V is then opened or closed so that the water level in the V falls or rises. Naturally, the pitch of the notes produced falls or rises with the water level. It probably does no harm to imagine that the ma eeryeh sounds something like blowing across the mouth of a partially filled soft-drink bottle.

The orchestra thus consisted of two bwikis, two musmars, and two ma eeryehs. A fourth instrument, not involved in the opening numbers, stood mysteriously silent, its musician looking straight up as the concert began.

It cannot be doubted that the orchestra was far more interesting to hear than to see. We absorbed ourselves for a time watching the musicians. The only conclusion we reached after half an hour's observation was that the ma eeryeh was played twice as fast as the musmar and the bwiki was played twice as fast as the ma eeryeh. We scanned up to Yendred and Na.

▪ MAY WE INTERRUPT?

THIS IS REALLY QUITE GOOD. THE MUSIC MAKES ME THINK OF DISTANT PLACES SO I ENJOY TALKING WITH MY EARTH FRIENDS AS I LISTEN.

▪ AS YOU KNOW, WE CANNOT HEAR THE MUSIC. THE MUSICIANS APPEAR TO BE READING FROM SHEETS IN FRONT OF THEM. AFTER THE CONCERT WE WOULD VERY MUCH APPRECIATE YOUR GOING DOWN AND BORROWING THE SHEET MUSIC FOR A WHILE.

DO YOU WANT ME TO READ IT TO YOU IN SOME WAY? I KNOW THE NOTES AND THEIR VIBRATIONS.

▪ THAT IS PERFECT. IS NA ENJOYING THE MUSIC?

SHE IS A MUSICIAN. SENSITIVE ALSO. SHE CAN HEAR THE SINGING OF THE STARS.

Even at this stage of their relationship I could predict that Yendred would have trouble saying goodbye to Na. As for the music, Yendred assured us that it was only necessary for him to get one score; everyone played exactly the same melody, each instrument playing twice as fast and one octave higher than the instrument below it.

After several pieces of this nature, the players paused for a long time. During the interval, the hitherto quiet musician stirred herself enough to pick up her score and study it briefly. We all came more or less simultaneously to the same realization as she began to check the tuning of the tall and complicated instrument beside her.

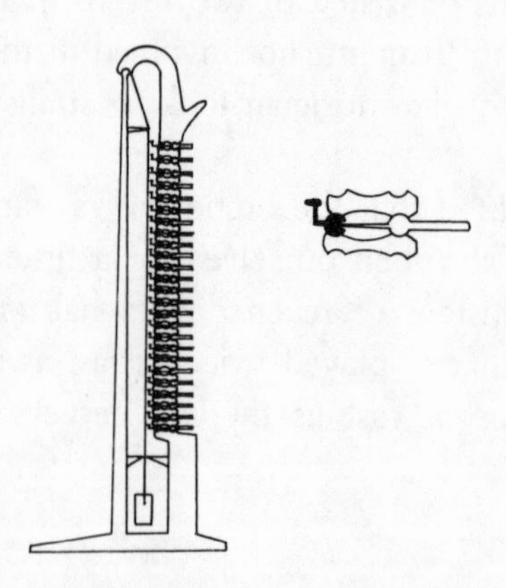

It was clearly a piano with one string and a vertical keyboard. A weight inside the instrument kept the string at the proper tension. When the musician struck any of the keys, a small hammer swung into the string, striking it and causing the upper portion to vibrate. A pair of dampers just above the weight kept the lower part silent. Vibrations of the string were communicated through a bridge to the soundboard, which was also the instrument's main structural support.

The "piano concerto" (or whatever it was) began, and we noticed that the pianist was careful not to strike the keys too forcefully. There was very little to prevent them from popping out of position and falling into a jumble of two-dimensional hardware at her feet.

We enjoyed the piece, visually at least, and learned from Yendred that this instrument had only lately been developed right there in Sema Rhublt and was quickly becoming a favorite of the concert-going public.

When at last the performance ended, Yendred and Na left with the audience in the western half and, at the first traffic pit they came to, entered to let the rest of the audience pass over them. Now Yendred and Na could return to the amphitheater and get the sheet music. Na, however, was puzzled why Yendred was so anxious to examine the score. He told her he wished to analyze it briefly. This seemed to impress her, for, as the two climbed over seat after seat down toward the orchestra, Na remarked on Yendred's intelligence and explained how nicely he would fit into the society of Sema Rhublt. Arriving at the stage, Yendred asked the musicians if he could borrow a page of their music and one of the bwiki players, with an almost careless wave of the arm, told Yendred he could keep it.

This was a stroke of luck because Yendred presently had very little time to relay this fragment of the score to us. Several days later he was able to transmit and explain the music at his leisure.

On the way back to Na's house Yendred paused in the almost incessant conversation with his new friend.

HAVE YOU LEARNED VERY MUCH TODAY?

Yes, indeed we had. It had been a most informative day. Well, then, said Yendred, it did not appear that there would be much more for us to learn. Perhaps it was as late on Earth as it was getting to be on Arde.

We took the hint and closed down the 2DWORLD program. Yendred would be asleep during most of next evening's contact period, so we

TEMPO FUGUE

The piece displayed below has been scored for three instruments in the key of C minor, this being the standard pitch closest to the note frequencies which Yendred gave us. Although the music seems to be in 4/4 time, we never learned if this was so. Accordingly, I have been hesitant to write that signature even though the bars and time-values imply it. The piece continues, of course, but these were the only notes appearing on the page Yendred obtained.

The music begins with the bass notes played by the musmar. The ma eeryeh then takes up the same sequence of notes at twice the speed in bar five and then, in bar seven, the bwiki enters with precisely the same sequence twice as fast again.

If this is a typical example, then Ardean music takes some getting used to for human ears. Although the melody is not too unearthly, the harmonies are thin and the musical ideas being expressed seem uncertain. Perhaps we might use a synthesizer some day to simulate the Ardean instruments, the brief glissandos at the onset of notes, and the echoing Ardean airspace, to form some notion of what Yendred and Na actually heard on the night of the concert.

agreed to wait until Tuesday. Since it was just past 1:30 A.M., most of us were grateful for the chance of a good night's sleep. Alice followed me to my office. "I think I'm going to stay for a while and work on my thesis."

It occurred to me suddenly that Alice had been sadly neglecting this area of her life in favor of our contacts.

"Hmm. I guess you'd better. Will you be on the machine?" I envied the physical resilience of the young.

Tuesday, July 15, 9:00 P.M.

By 9:15 we had turned on the computer and got 2DWORLD up and running. Wobbling into focus was an enormous rectangular structure divided down the middle by a post with complex internal details. This was evidently the gondola of the freight balloon which would carry Yendred to Dahl Radam.

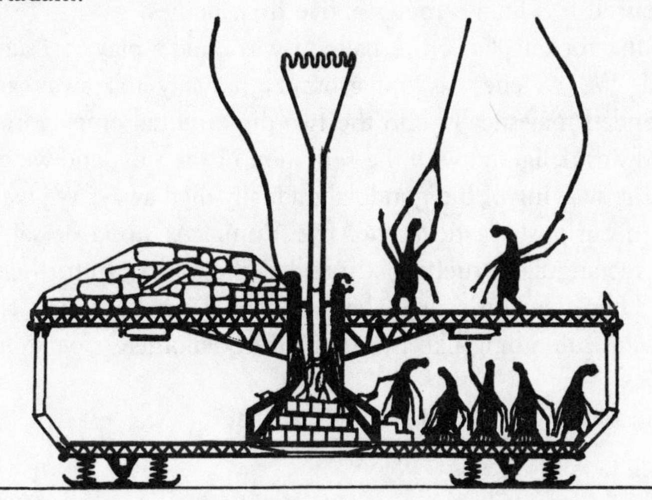

Yendred was already aboard, being one of the three passengers in the eastern compartment. The western half of the gondola was empty, but its roof was piled high with goods which the balloon would carry to a special depot on Dahl Radam. The trip would take two afternoons, with a full day's stopover at a small town midway between Sema Rhublt and Dahl Radam.

There were five Punizlans in all standing or sitting in the eastern compartment. At the eastern window the balloon captain watched the ground crew cast off restraining ropes. Earlier he had given the order for his crewman to throw a switch on a panel affixed to the central post. This had cut in a series of powerful batteries beneath the floor of the gondola to activate a heater mounted on a long pole extending above the central post. From time to time the ground crew had released and caught again the balloon's skin, trapping more air beneath it as the "bag" inflated near its top with hot air from the heater. Before the last rope was cast off, a balloonport employee on the roof of the craft had inserted the free edge of the bag under a clamp on the post and levered it shut.

And now the gondola shuddered slightly as it began to skid eastward along the sand of the balloonport. The ground crew retreated cautiously from the balloon, no doubt hoping it would soon lift. They lay flat as the gondola almost painfully and reluctantly hove into the air, swinging massively as it passed less than a meter above their bodies.

Unlike the rocket plane, the balloon was child's play to track from our terminal. We watched as the ground gradually fell away and the gondola ascended majestically into the two-dimensional empyrean.

Yendred was delighted with the sensation of the ride, and we chatted with him as the swaying of the gondola gradually died away. We were left, finally, with a curiously static scene: five Punizlans stood or sat in the middle of a rectangular structure suspended motionless from the top of our screen. There was no sense whatever that the balloon was clipping along by now at something like twenty or thirty kilometers per hour.

ALICE WOULD ENJOY THIS MORE THAN THE ROCKET.

▪ ALICE IS NOT HERE.

I AM GOING TO ASK THE CAPTAIN TO LET ME LOOK OUT THE WINDOW.

It must be said that windows are a rare treat for Ardeans. Their houses, being underground, have no need of windows, and there had been no windows on the rocket plane except for the pilot's.

The captain, obligingly enough, allowed Yendred to come to the window at the cost of having the captain walk over him. The scene sounded captivating.

SUCH VASTNESS BELOW. THE GROUND IS SIMULTANEOUSLY TINY AND HUGE. ALL SHADES OF BROWN AND SO FAR DOWN. HERE AND THERE ARE GREEN DUST PATCHES. PLANTS, I SUPPOSE.

He strained his eyes eastward for a glimpse of Dahl Radam, the enormous plateau of Ajem Kollosh which we also looked forward to seeing for the first time. There would lie the answers to his questions, he told us. There he would find the object of his search.

In the midst of his spiritual enthusiasm, he felt the captain's hand tapping on his neck. He swung his head 180 degrees away from the window to find the captain staring at him. That would be enough.

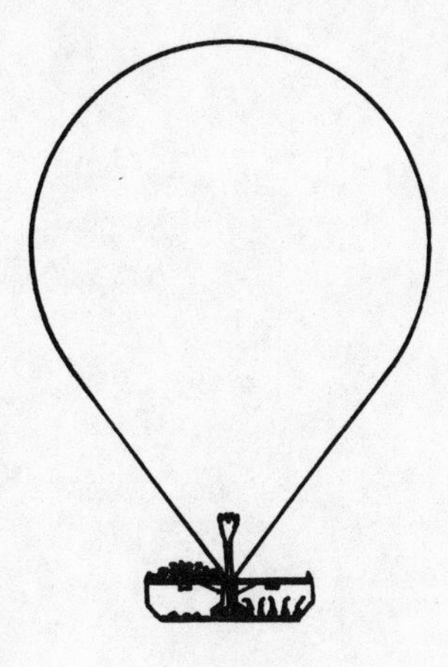

9

High on Dahl Radam

Friday, July 18, 9:00 P.M.

Although we had missed the stopover between Sema Rhublt and Dahl Radam, we were able to renew contact with Yendred nearly two hours before the freight balloon landed at the small station called Mkien Tarj, halfway up the gentle slope of Dahl Radam. We found Yendred in a state of earnest conversation with a fellow passenger and so contented ourselves for a time with a long look at the balloon, gondola, and occupants. This turned out to be a lucky thing for us: not three minutes into the contact, a strange creature flew directly under the balloon, traveling in the opposite direction. It was less than a meter long and looked for all the world like a snake! It puzzled us that a snake could fly until we realized that there was no reason why a snakelike creature shouldn't fly in the atmosphere of Arde: it could not fall quickly because of the air trapped beneath its body. It could even make forward progress, flying in effect, by catching pockets of air in its curves and working these toward its rear. We left the balloon and followed the snake for a short time, recording what we could of it until contact was threatened by our distance from Yendred.

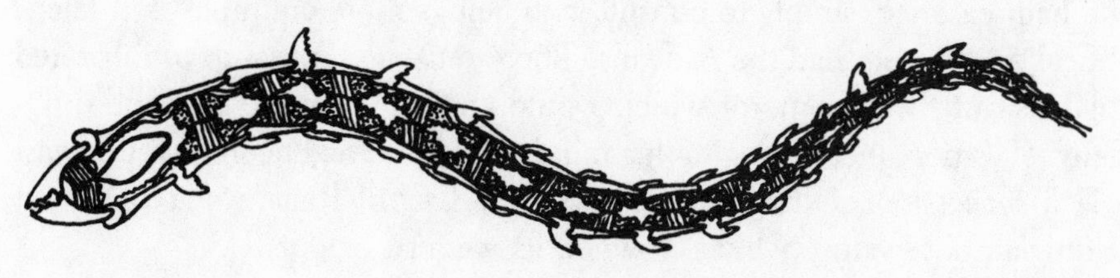

When we found the balloon again, Yendred was still talking with his fellow traveler.

▪ EXCUSE US PLEASE. WE ARE BACK. WHAT WAS THAT CREATURE THAT FLEW BY?

IF IT WAS LONG AND THIN IT WAS A BES SALLUR.

▪ DO YOU HAVE OTHER ANIMALS THAT FLY?

LET US TALK LATER. I AM SPEAKING WITH THE ONE CALLED TBA KRYD.

We excused ourselves. What was Tba Kryd doing on the balloon? Previously, he had been heading west from Maj Nunblt. Our failure to recognize him among the passengers was perhaps forgivable: all Ardeans look more or less alike to Earth people.

▪ EAVESDROP.

AND ASTRONOMY THE ULTIMATE ANSWERS MAY HOLD. IF ONLY INTELLIGENT BEINGS THERE ARE IT THE MIND ASTOUNDS. EVEN NOW PERHAPS THEY US SIGNALS SEND. I TO MY PLAN THE ASTRONOMERS WILL AGREE HOPE.

As we observed this conversation it occurred to me that Yendred might well be enjoying a certain delicious smugness, betrayed by an almost patronizing enthusiasm for the discussion about intelligent beings beyond Arde. If this was the case, all conversations with Tba Kryd would have to be monitored carefully.

Yendred had not recognized Tba Kryd during the earlier portion of the flight because another passenger had sat between them. Apparently, while still traveling to Is Felbt, Tba Kryd had changed his mind about pressing for changes in the Punizlan political system. How much easier, he had reasoned, simply to consult with beings more culturally and scientifically advanced than the Ardeans. There must be thousands of inhabited planets out there, some of which would know the solutions to the problems of Arde. Since changing his mind, Tba Kryd had been traveling east via a succession of freight balloons, toward Dahl Radam and the great Punizlan observatory which crowned its western heights.

DO YOU THINK IT WOULD HELP TBA KRYD IF HE COULD TALK TO EARTH PEOPLE?

▪ I THINK IT WOULD BLOW HIS MIND.

I asked Lambert if he would mind my taking over at this point.

▪ DEWDNEY HERE. OF COURSE WE COULDN'T TALK TO TBA KRYD DIRECTLY AND AS WE HAVE MENTIONED BEFORE IT WOULDN'T EVEN BE A GOOD IDEA TO TALK TO HIM THROUGH YOU.

YES I KNOW. BUT DO YOU HAVE ANY ANSWERS FOR HIM?

▪ PROBABLY NOT. EARTH AND ARDE ARE MUCH TOO DIFFERENT.

Before long we saw the balloon crewman who operated the power switch turning it off for longer and longer periods of time, allowing the air in the vast balloon overhead gradually to cool and shrink. They were coming in for a landing. The captain strained forward, looking through the clear jebb window. The crewman threw the switch on again as the ground came up to meet the skids on the bottom of the gondola. With a sharp thump and a long slide, it finally came to a lurching halt, the balloon bobbling into view on the upper right half of the screen.

Here was a much different place than any other on Arde! The soil was coarser and filled with sharp, irregular stones and boulders. Moreover, the whole surface sloped up and to the right about 10 degrees. This was indeed the slope of Dahl Radam.

Some Ardeans walked into view from the station, which the balloon had overflown slightly. They climbed over the empty half of the gondola and released the balloon skin from the clamp, flinging it clear of the heater. Next, they disengaged the heater itself so that the freight could be unloaded and so that the passengers could climb over the gondola on their way to the station.

Yendred followed Tba Kryd to the station, passing over the few homes and small warehouse of Mkien Tarj. The station was a four-storey structure whose uppermost level contained a waiting room for balloon passengers. Word had come by radio of a heavy rainfall on the upper slope, and a nearby resident whose duty it was to ring a gong at such times now did so. Soon a river would be upon them.

Yendred busied himself by descending to the next level, where there was a small store. He spent almost all of his remaining money on another travel balloon and some food supplies. Apparently, he could not afford to travel any farther by freight balloon. When he returned to the uppermost level, he was confronted by Tba Kryd.

YOU HOW MUCH MONEY HAVE?

TWELVE JEBB ONLY.

AH THAT EXCELLENT IS. I PRECISELY THAT MUCH EXTRA FOR MY LAST TICKET NEED.

BUT HOW YOU RETURN WILL?

I SEVERAL OF THE ASTRONOMERS KNOW. THEY VERY KIND ARE. BESIDES I MUCH CREDIT HAVE. I YOU OF COURSE WILL REPAY.

Suddenly, everyone in the waiting room cringed visibly, as if at a sudden noise. Scanning up to the surface, we found an astonishing flood of water passing over the entrance. Stones, boulders, and sand, all mixed together, tumbled, rolled, and rebounded within it. By the time we had recovered from our shock at this sudden flood, it was virtually over, like a freight train on an overpass. We all thought of the freight balloon simultaneously and scanned upstream until we found the gondola half buried on its eastern, upslope side in rocks and sand.

We thought it rather careless of the Punizlans not to provide an underground storage shed for the gondola. At first it seemed to us an almost hopeless task to free the enormous box from its rocky tomb. On closer examination, however, it became clear that a few shovel scoops would clear off the roof and a brief excavation on its eastern side would make the passenger compartment once more accessible. Again, we had been caught up in three-dimensional thought patterns, imagining this to be merely the cross section of a much bigger job.

When the flood had finally subsided, it was time for the passengers to bed down for the night at Mkien Tarj. Yendred, no longer a ticketed passenger on the Punizlan Balloon Service, was not allowed to sleep at the way station. He was allowed, however, to spend the night in the gondola. Even at this height the nights were very cold. Given a small portable electric heater, he made his way out of the station onto the now starlit surface of Dahl Radam. From far away he could hear the thunder of the river, already many kilometers downslope.

Yendred made his way wearily over the flood debris, unlatched the western window, and clambered into the compartment. He plugged in the heater and settled down for the night on two seats after filling the space between them with a long rope from a box near the control panel.

▪ ARE YOU WARM ENOUGH?

I AM OKAY AS YOU SAY. I AM MORE COMFORTABLE ALONE ANYWAY.

▪ TELL US ABOUT THE BES SALLUR.

We did not find out a great deal about this fabulous creature before Yendred's sentences began to trail off, as though our printer were caught up in the contagion of sleep.

Bes Sallurs live on the slopes of Dahl Radam and are occasionally found even on its heights. By day they fly high above the surface, their cruel crystalline eyes searching for the tiniest hint of movement on the ground; perhaps a small burrowing creature has emerged into the light to feed or freshen its gills. Seeing one, they wait until it has fully emerged and then swoop down from the sky, landing with a terrifying burbling sound to trap the hapless thing between head and tail, enclosing it in a pocket of death from which it has no time to burrow. Tucking its head beneath its body, the Bes Sallur bites its victim into pieces, swallowing them one by one.

By night these flying snakes crawl beneath leaves or excavate loose gravel to cover themselves with. In the cold they become inactive, entering a state like hibernation.

Each spring the Bes Sallurs perform a curious mating dance in which one sex assembles upslope and the other down. Within each group there is a furious struggle to see which shall be first in a great procession toward the other group. When the two processions meet, the individual Bes Sallurs pair up in the order of their encounter, crawling over already mating couples until they meet their intended head on.

Saturday, July 14, 9:00 P.M.

Yendred had already said goodbye to Tba Kryd, who would continue on, the next day, to the observatory. With difficulty Yendred struggled up the side of the gondola, over the debris-laden deck, and onto the layer of rocks deposited by the river on the gondola's eastern side. He was alone and penniless, and Lambert was in an unusually solicitous mood.

▪ ARE YOU SURE YOU WANT TO CONTINUE? PERHAPS IF YOU WAITED AT MKIEN TARJ YOU WOULD GET LUCKY. PERHAPS YOU WOULD GET A FREE RIDE.

NO. IT IS BETTER THIS WAY. WHEN I AM ALONE AND TRAVELING WITH A PURPOSE A CERTAIN PLEASURE COMES OVER ME. THIS IS HOW THINGS WERE MEANT TO BE.

There is no arguing with a fatalist.

We followed Yendred's progress up the gradual slope. Clearly, the tough going would soon exhaust him. Why didn't he set up his travel balloon? Yendred said that this would interfere with his visit to a certain monument, which he could now see looming in the distance to the east. On the way Yendred told us what the monument was and began a long account of Ardean history.

The monument was all that remained of an enormous fort constructed hundreds of years ago by a Punizlan army to ward off a Vanizlan attack.

Long before that, Vanizla was a civilized land and Punizla a waste inhabited by disorganized and constantly warring tribes of barbaric Ardeans. Early in Vanizlan history, a great leader called Ajeb brought a Vanizlan army across Dahl Radam to conquer and subdue the tribes of Punizla. In this he was successful, and founded a single, unified empire which spanned the whole of Ajem Kollosh. The empire flourished for a thousand years. At that time Ardeans believed that all things, great and small, were inhabited by spirits. The greatest spirits inhabited the greatest things, so it was only natural that the spirit of Arde itself, and the spirits of Shems and Nagas, should receive the most worship. A priestly class grew up around these beliefs and became very powerful, mediating between the intangible spirits and the tangible Ardeans. Decay of this empire began with rebellion in Punizla and ended with the rise of a prophet called Amada in Vanizla itself. In one sweep the earlier structure of religion and state was erased and a second empire began.

Although portions of Punizla were reconquered by the renewed Vanizla, the greater part remained uneasily independent, governed by a pan-tribal council. At this time many wars were fought, mostly in eastern Punizla. Such conflicts were characterized by the meeting of two armies of soldiers. Each soldier was equipped with weapons of the time. One of these was a sword-and-shield combination, each wielded by a hand on one side of the body. Another was a double-handled spear which could also be used as a shield. Other defensive armor was obviously unnecessary.

On Earth, primitive conflicts of this sort were generally preceded by single combat between champions from each side. On Arde, a general melee was, of course, impossible, and the entire battle consisted of a succession of single combats between whichever two soldiers stood at the head of their respective armies. The best warriors were placed at the head of the army and the weakest at the tail. By the time the battle had been in progress for the better part of a day, there would be a heap of dead bodies upon which the current pair of warriors fought.

Yendred told us the story of a famous Punizlan warrior of this time, Rachytl by name. He had been placed at the very end of a small contingent protecting the village of Rajm Kratk from a marauding Vanizlan force. He is commemorated in a famous epic poem in the ancient Punizlan language. Yendred could only remember the first few lines. It came through the translation process rather strangely.

OH YOU RAJM KRATK OF NSANL
WEAKEST RACHYTL WAS.
OF ALL OH YOU.
UNDER SHEMS IN RAJM KRATK
WHILE ALL OF YOU OH YOU DIED
ONE BY ONE. . . .

The long and short of it was that Rachytl spent an hour waiting under a hot "sun" hearing but not seeing his comrades die, one by one. When his turn came, he was hoping against hope that there might be at most one Vanizlan left to deal with. But fighting for his home village gave him the courage and strength of ten Nsana. A sword thrust killed the first opponent, but a second appeared behind. This battle (and Yendred's account of it) went on for some time. Finally, after working his way through something like a hundred Vanizlan warriors, Rachytl stood alone, sword in one hand, shield in another, atop an enormous pile of dead bodies. He died from his wounds, not surprisingly.

After a conflict which lasted hundreds of years, the Punizlans gradually pushed the Vanizlans back. They even succeeded in taking Dahl Radam. At the eastern head of the plateau, where it begins to slope downward toward Vanizla, the Punizlan leader conceived a horrible plan for wiping out the Vanizlans once and for all.

The Punizlans built a great dam on the slope, made from huge blocks of jebb quarried from Dahl Radam itself. They brought blocks from the quarry on rollers and ferried them across the lake which was slowly rising behind the dam.

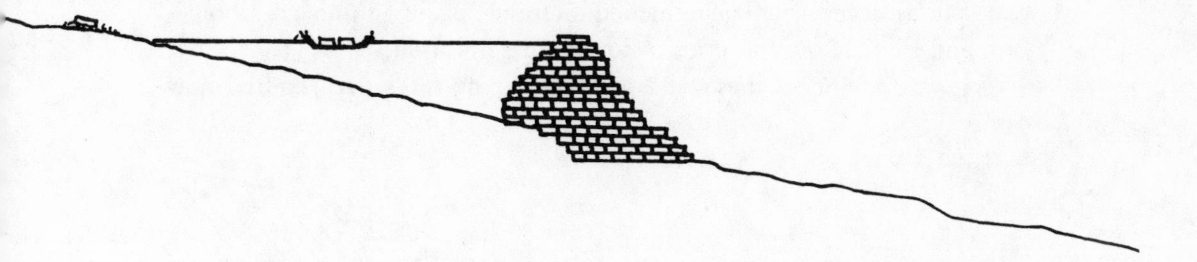

Such a project might strike Earth people as a rather large one, but here again we realized that a two-dimensional dam requires far fewer blocks than a three-dimensional one of comparable height.

The whole idea of the Punizlan project was to remove a few key blocks from the base of the dam when the lake had risen close to the top of the dam. The resulting flood would surely drown every Vanizlan from Dahl Radam to Fiddib Har.

One day the dam was demolished and a terrifying flood swept down on Vanizla. Only an enormous, sloping counter-dam, constructed in secrecy by the Vanizlans a hundred kilometers to the east, prevented their extinction. From chambers beneath that dam, a force of 1000 Vanizlan soldiers emerged and deployed buried boats on the floodwaters. Landing on the western shore, they fought their way to the height of Dahl Radam, discovering in the process some stones from their sacred shrine which had been used in the dam. In their fury, it took them only a few weeks to recapture all of Dahl Radam and rebuild the shrine. For many years the

Vanizlans held Punizla at bay while they worked desperately to restore an agriculture severely disrupted by the flood.

Yendred seems to have skipped a long stretch of history at this point, mentioning only that Punizla had developed the beginnings of its scientific and technological character not long after this. A great flowering of learning took place: Punizlan explorers completed the map of Arde, even surprising the Vanizlans by landing on their eastern shore. Medicine, science, and the arts all flourished. The Punizlans subjected everything in their experience to a merciless rational scrutiny. They even modified their version of the Vanizlan religion: they discarded doctrine that made no sense, and dropped some of the more strenuous observances.

This was also the time when a new series of conflicts between the two nations began. Vanizla launched a crusade against Punizlan rationalism, and Punizla built a series of forts to defend itself.

It was to one of these, the only remaining fort, that Yendred now came.

Forts, like dams, are easy to construct on Arde. This one was badly eroded near the top, at least ten courses of stones missing. Behind the fort, over the last two centuries an enormous body of rock, soil, and debris had built up. This resulted, to Yendred's eyes, in a steep stairway leading to a new slope on the heights of western Dahl Radam.

Here it was that the Punizlan army had once camped, fighting off the Vanizlan crusaders who tried to scale the wall. It was impressive to realize what the lack of a third dimension could do to military tactics: not only did a fort require only a single wall, but it could protect a vast stretch of territory.

By the time this second great period of conflict had ended, Punizla had resumed its original scientific and artistic flowering at an ever-increasing pace. But new, internal developments threatened the peace. The Great Planner Revolt took the four eastern regions out of Punizla and a new nation was formed. It was called Laklb.

Yendred reached the top of the fort and set up his travel balloon. He filled it to just the right size, secured his belongings in the carrier bag, and, with a gentle leap, was airborne—at least for a few meters. As he bounded with an almost kangaroo-like pace across the surface under the gradually freshening westerly, he spoke of the intense hatred between Punizla and Laklb, of new developments in weapons, including explosives and attack rockets, and of the devastating conflict which had finally erupted.

Ninety years before the present time, a Vanizlan cargo balloon had developed a break in its skin and had plummeted to the ground at the site of a volleyball match in Laklb. Two bombs, which just happened to be part of the cargo, exploded and killed most of the players and spectators. The Planner government interpreted this as a preemptive strike from the west and unleashed its invasion force.

▪ WHAT HAPPENED THEN?

THE WAR LASTED FOUR YEARS. THOUSANDS OF NSANA WERE KILLED. FOR US THAT IS MANY MANY.

▪ WHO WON?

PUNIZLA OF THE FOUR REGIONS WON FINALLY BECAUSE OF HELP FROM VANIZLA. THEY ATTACKED LAKLB FROM THE EAST AND SUPPLIED US WITH FOOD BY SEA. EXCUSE ME. THERE ARE SOME NSANA AHEAD.

There were two Punizlan government seed-planters. The first walked along poking holes in the ground with a rod and the second placed seeds in each hole. Here was the origin of many of the edible plants which flourished in Punizla. Yendred bounded over the pair, shouting the customary Punizlan greeting from the air.

Our contact period was now nearly at an end. We were about to close down 2DWORLD when there came a knock at the laboratory door. We all froze. I decided to go in person, mentally preparing a story as I went.

"Alice!"

"Hi, I'm back."

She had an air of apology and repentance, although I hadn't a clue what she would be feeling sorry about. Before these last two sessions, she had never missed a one; we assumed that she had been either catching up on her sleep or working on her own research. Her eyes looked tired.

"We missed you, but this isn't a great time for you to come. We're about to close up shop."

She knew. She just wondered if Yendred had left Sema Rhublt yet. I filled her in quickly on the freight balloon, the depot, the flood, and the war stories. She took over at the terminal to sign off.

▪ ALICE HERE YNDRD. WE HAVE TO GO NOW. NEXT CONTACT IN [Alice paused.] 21 ARDE-HOURS.

WHERE HAVE YOU BEEN?

Alice hesitated and then typed:

▪ WORKING ON MY THESIS.

I MISSED YOU.

▪ GOOD NIGHT, YNDRD.

Alice smiled somewhat sheepishly as she terminated the 2DWORLD program.

The next two possible contact periods coincided with unfruitful times on Arde (Yendred would be sleeping most of the time), so we gave ourselves two days to catch up on our own sleep and regular duties.

Tuesday, July 22, 4:00 P.M.

Once our line with Arde was established, there ensued the most bewildering session we had ever had. Yendred was in some kind of trouble.

HELLO. I CAN SEE YOU ALL NOW. IT IS NICE OF YOU TO VISIT.
DEWDNEY YOU LOOK TOTALLY WEIRD. PLEASE. STAY AWAY.

▪ YNDRD WHAT IS THE MATTER?

NOTHING. JUST KEEP YOUR DISTANCE.

▪ WE ARE NOT THERE.

Yendred stood alone on the rocky surface of Dahl Radam. The grade was a bit more level now and we judged that he must be close to the top. He held all four arms out fending away invisible things and craning his head left and right.

THEN WHAT ARE ALL THESE CREATURES?

▪ WE DON'T SEE ANY CREATURES.

This strange behavior continued for three or four hours. Occasionally we could get Yendred to take a few steps of his journey, but then he would claim that a pit had opened at his feet and that we were merely trying to trick him.

Gradually, however, his statements and gestures developed a new character.

LOOK. JUST LOOK. HERE ALL AROUND ME IS THE BEAUTY OF THE
PRESENCE. THE SOURCE OF BEAUTY.

As usual, we saw nothing, but kept silent. His arm movements became slow and ecstatic. It was Chan who suggested the best explanation for Yendred's strange behavior.

"Oxygen deprivation—or maybe I should say 'hrabx deprivation'! His brain is starved for oxygen, like."

This certainly made sense. Yendred had only just attained—or was now attaining—the full altitude of Dahl Radam. The air would certainly be thinner and thus would contain less of the life-giving gas. Perhaps we could get him to flap his arms to aerate his gills.

▪ YNDRD. PRETEND YOU ARE A BIRD. FLY LIKE A BIRD.

"Arde doesn't have any birds, you idiot," said Lambert to Alice.

▪ MAKE YOUR ARMS GO UP AND DOWN.

But it was useless. Yendred continued to rave, but at least he was now walking, and in the right direction. We decided against advising him to use his travel balloon and to let him walk. It was now, according to our calculations, well past sunset on Dahl Radam. We had no idea how far away the observatory was or whether Yendred was competent even to make himself a meal or camp for the night.

THE BEAUTY IS COLD AND MAJESTIC.

▪ ARE YOU COLD? IT MUST BE FREEZING THERE.

YES. FROZEN BEAUTY. THE BEAUTY OF DEATH.

▪ DON'T SAY THAT! KEEP WALKING. WHY NOT RUN FOR A WHILE.

RUN WHERE? TO THE LIGHT IN THE EAST? TO THE DARK IN THE WEST?

▪ DO YOU SEE A LIGHT?

YES. IT IS LIKE A STAR BUT BRIGHTER.

We hoped fervently that this was a light on the observatory. Presently we found ourselves chanting, "Go, Yendred, go!" He walked unsteadily, obviously impressed by our concern but given, at odd moments, to stopping and looking straight up.

THE STARS ARE LIKE. OH THE STARS ARE LIKE. THE STARS ARE UNSPEAKABLE UNTELLABLE UNBELIEVABLE.

▪ KEEP WALKING.

Suddenly another Ardean came onto the screen.

▪ EAVESDROP

YOU YNDRD ARE? WE YOU HAVE BEEN EXPECTING.

The Ardean carried an electric lantern. He introduced himself as an astronomer attached to the observatory, which was not far ahead. He took Yendred's hand to steady him and, as they walked, he explained to his largely uncomprehending charge that he was suffering from a well-known sickness which comes upon anyone who ascends Dahl Radam too quickly. It was called the "Deadly Joy." Luckily, some part of Yendred's mind had made him keep hold of his string bag. Luckily also, I suppose, Yendred had us.

The observatory was a massive and impressive installation. The astronomer took Yendred down to a guest room where he could be treated for the Deadly Joy by being made to drink a strong herbal brew. We scanned up to examine and record the observatory as a whole.

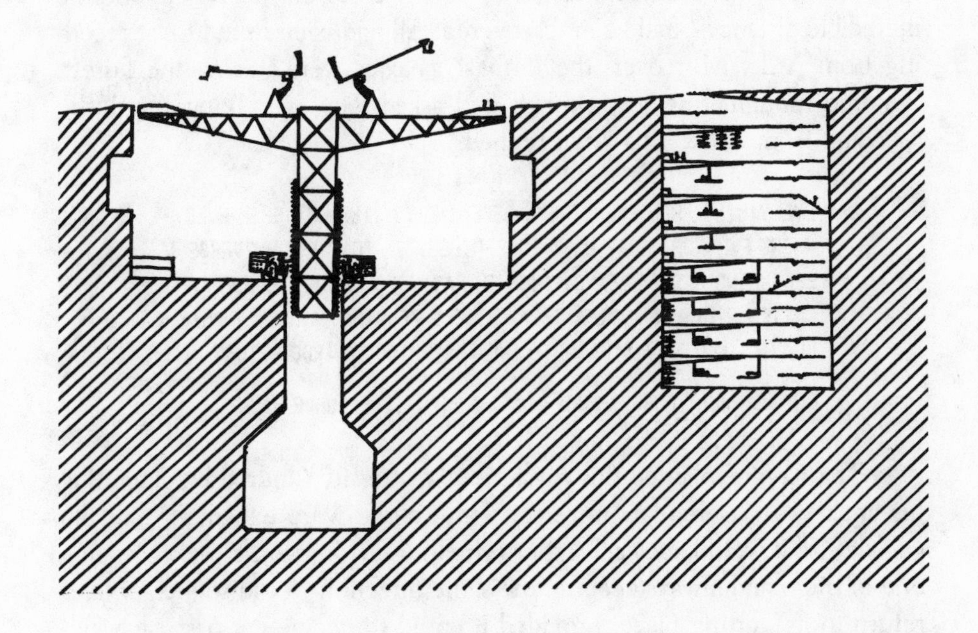

Two enormous telescopes were mounted on a platform which could be raised or lowered by two steam engines near its base. One telescope, for viewing the western half of the sky, was not in use. An astronomer sat up near the focus of the other one, photographing some distant galaxy off in the east.

Back in his room, Yendred was now somewhat recovered. Tba Kryd was there.

I TOO OF THE DEADLY JOY A TASTE HAD.

However, Tba Kryd had been treated immediately when the freight balloon had landed. Consequently he had been spared its mental anguish and had recovered very quickly. As he talked he moved his arms rhythmically up and down, advising Yendred to do the same. With the enthusiasm we had come to expect of the good philosopher, he described the observatory and its work to Yendred.

The entire circle of the night sky was being photographed at very high resolution. Already, hundreds of new galaxies, some of them at quite incredible distances, had been discovered. All had been found to be receding from Arde. Moreover, the farthest galaxies were fleeing the fastest. Yendred thought this very strange and asked why everything should be receding from Arde. Tba Kryd replied,

EVERYTHING FROM EVERYTHING RECEDING IS. THE UNIVERSE EVERYWHERE EXPANDS. THAT MOST INCREDIBLE IS. THAT THE UNIVERSE A BEGINNING HAD IT MEANS. ALSO THE ASTRONOMERS ME TELL THAT THE UNIVERSE ITSELF LIKE A GREAT CIRCLE REJOINS. A ROCKET SHIP ARDE LEAVING AND IN A STRAIGHT LINE TRAVELING TO ARDE RETURN WILL.

BUT HOW WHAT FOREVER GOES ON CAN ON ITSELF CURVE BACK?

Tba Kryd explained this mysterious matter to Yendred by extending his analogy of a circular universe: if the circle were expanding, all the galaxies on it would see each other retreating. Moreover, a rocket leaving one of them and always keeping the same direction would sooner or later return to its starting place, provided it could somehow get past each galaxy. Yendred pretended to ponder this but turned to us for help.

IS THIS TRUE?

▪ YES. IT WOULD SEEM SO. THE SAME SORT OF THING MAY BE HAPPENING WITH OUR OWN UNIVERSE.

Assuming that the Punizlan astronomers were correct about the shape of the Planiverse, that it is in fact the simplest two-dimensional analogue of a circle, I am tempted to imagine the entire Planiverse as a vast sphere sprinkled with innumerable galaxies, these arranged into clusters of galaxies. Perhaps the clusters themselves are so arranged.

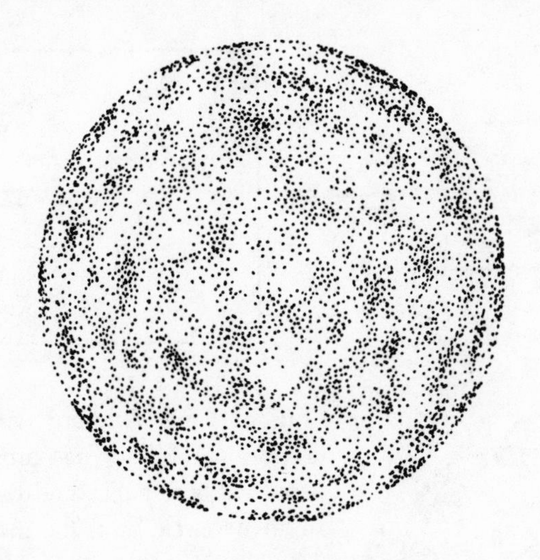

The Planiverse is thus a sort of cosmic balloon whose infinitesimal skin contains all that can possibly exist for the Ardeans: galaxies, stars, planets, homes, and Ardeans themselves.

It was the galaxies which provided Punizlan astronomers with their clue to the expansion of the Planiverse.

What do these galaxies look like? According to the Punizlan astronomers, each contains millions (not billions) of stars and each star is a disk of two-dimensional matter undergoing a violent but long-lasting reaction, radiating light outward from its rim in all directions.

Unfortunately, the Punizlan astronomers cannot tell if these galaxies have a spiral structure or not; seen edge-on, so to speak, it would be impossible to detect spiral arms in a Planiversal galaxy. Nevertheless, it becomes easy to appreciate the size of the Planiverse if we contemplate the possible appearance of one of its galaxies and imagine for a moment that one of those tiny dots is Shems itself.

THE EXPANDING PLANIVERSE

The Punizlan astronomers have observed what is probably only a small patch of the Planiverse and have noticed that all the galaxies are receding from them. They have concluded that this entire patch is expanding.

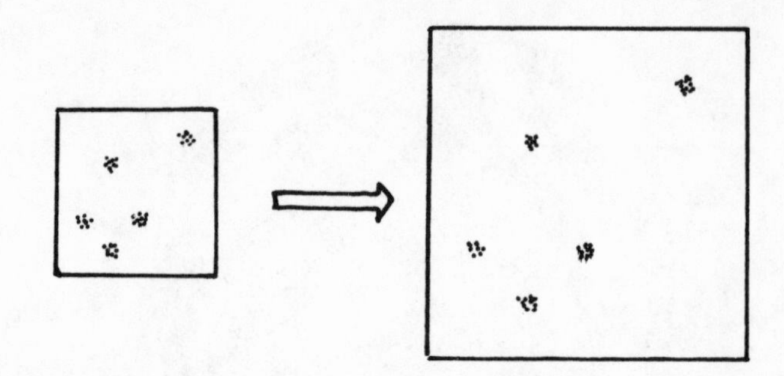

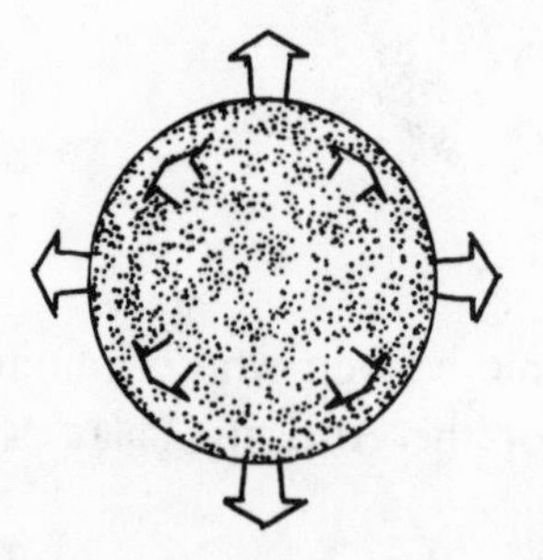

But of what structure does this patch form a part? They had probably considered various shapes and concluded that a "hypercircle" was the most likely possibility. By this term they mean a "sphere," even though they cannot imagine its appearance. The "hypercircle" is expanding, like a balloon.

Undoubtedly, any of the conceivable shapes is beyond the Ardean imagination to grasp. They may have considered analogies between the Planiverse and a straight line or a circle. Assuming that the Planiverse has no edge, how could the analogue of a straight line ever have been produced from a big bang?

Earth scientists are at present uncertain whether our own universe is a "hypersphere" or has some other shape. They are also uncertain about whether it will ever collapse again. But on the latter point, the Punizlan astronomers have no difficulty: sooner or later the tireless inverse linear law will pull all Planiversal matter back together again in one cataclysmic smash!

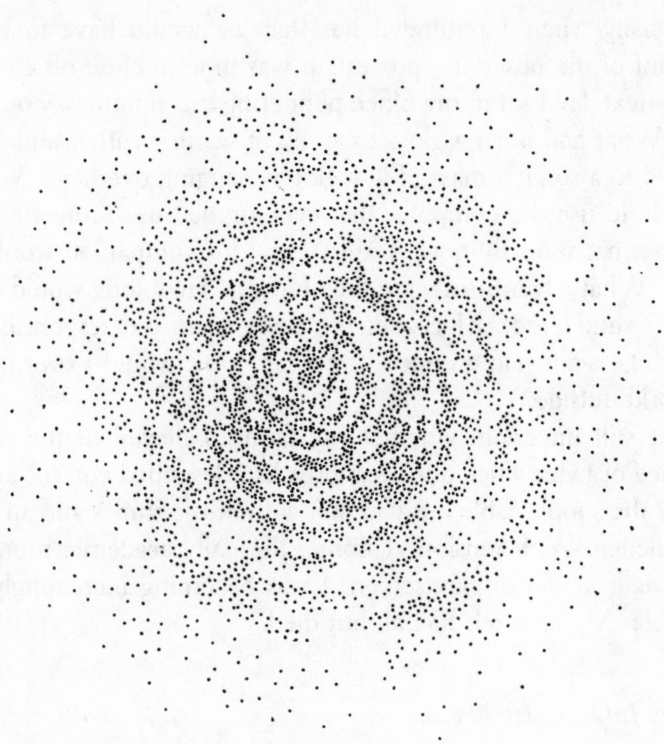

Turning back to Tba Kryd again, Yendred asked him if the astronomers had been interested in his idea of contacting other intelligent civilizations. It would appear that the idea had little appeal to them, and one had even scoffed at him. However, Tba Kryd had a completely new plan of action for guiding the affairs of Arde.

WE A NEW CIVILIZATION MUST CREATE. PUNIZLA AND VANIZLA ONE NATION MUST BECOME. I WITH YOU TO VANIZLA WILL GO.

"Oh, no!"

On previous occasions, Alice had made no secret of her feelings about some of the beings that Yendred encountered. She had once characterized Yendred's uncle as a "jerk" and had referred to Tba Kryd as "soft-headed." This time there was nothing beyond that small despairing cry, especially when I reminded her that we would have to keep our feelings out of the observing process. It was time to close off contact.

The next day I sat in my office pondering the gloomy sky outside my window. What had been a glorious week of warm weather and sunshine had turned to a sodden mass of low clouds hanging overhead. We would not be able to use the computer that evening because someone else had booked the machine. By now, Yendred and his companion would be on their way. What adventures were they having? How long would the contact last? I shuddered as I considered once again the responsibility this whole strange affair had dumped in my lap. How would I ever present it to the world outside?

I was still uncertain whether my family believed in the reality of Arde. Once or twice since our vacation I had described current events on Arde over the supper table but had met with brittle smiles and an uncomfortable silence. With tension at home, the daily academic routine, and all-night vigils at the display screen, I was becoming increasingly jumpy and irritable. My research had all but died.

Thursday, July 24, 10:00 P.M.

When contact was once again established, we found Yendred and Tba Kryd still at the observatory. It was morning and they sat in the upper level of the astronomy building, their pack beside them. The telescope platform had been lowered into the ground and a cable had been stretched over the pit as a bridge. There was nothing to do but wait.

On the lower levels two of the astronomers were sleeping. Another was busy developing and studying plates taken on the previous night, and two others sat at a table playing some kind of game. We watched the film-developing process for a few minutes but found nothing very remarkable, and so switched our attention to the game.

It is quite an interesting problem to watch a game with which one is totally unfamiliar and try to guess what the rules are. The two astronomers took turns placing pieces of jebb on a board which had been marked off into eleven segments, each having room for only one piece. A long period of thought preceded each move, although sometimes a player would put down his stone and pick up some others.

Later that day Yendred informed us that this was a very popular Punizlan board game called "Alak." One side has white pieces and the other black. If a contiguous set of one color is surrounded (at both ends) by two pieces of the other color, then that player gets to capture all the pieces thus surrounded.

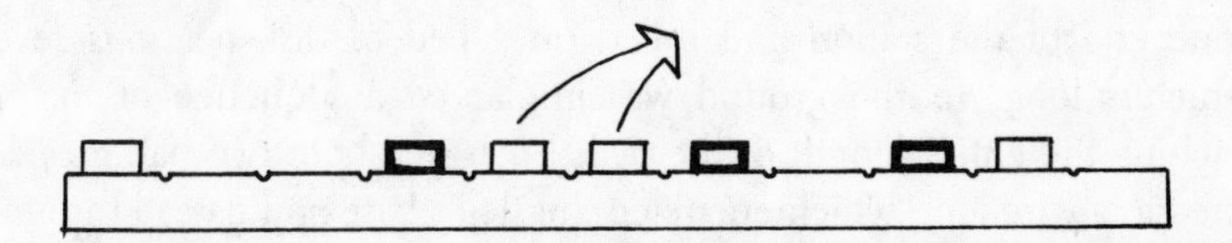

By now the wind had come up and it was time to set out. Three of the astronomers accompanied Yendred and Tba Kryd to the surface. Yendred had to inflate the travel balloon somewhat more fully in order to carry both him and his companion. Tba Kryd took hold of an upper hanger on the rope and Yendred a lower one. The astronomers watched as Yendred launched himself into the wind.

After a few hundred meters, Yendred decided that their hops were a bit short and injected some more gas into the balloon. Yendred was on the opposite side of the rope from Tba Kryd and was able to give us his full attention. Would he continue the history which had been interrupted several days ago?

I HAVE ALREADY FINISHED.

▪ BUT WHAT HAPPENED AFTER THE VICTORY OVER LAKLB?

THE POPULATION OF LAKLB WAS DISPERSED AND SOME OF THEIR CITIES WERE RESETTLED WITH PUNIZLANS. SEMA RHUBLT IS SUCH A PLACE.

▪ DO YOU THINK THE PLANNERS WILL REVOLT AGAIN?

PERHAPS. BUT THREE GENERATIONS HAVE PASSED. OF COURSE, THE PLANNER PHILOSOPHY HAS HAD GREAT INFLUENCE IN PUNIZLA. AND THE PLANNER PARTY IS STRONG. THAT MAY BE ENOUGH FOR THEM.

▪ WHAT ABOUT VANIZLA? WHY IS THERE SO MUCH BAD FEELING ABOUT VANIZLA?

THAT PUZZLES ME. IT HAS BEEN GETTING SLOWLY WORSE. PERHAPS WE

PUNIZLANS ARE ANGRY BECAUSE THE VANIZLANS WILL NOT ADOPT OUR CIVILIZATION.

Yendred and Tba Kryd now came to the last outpost of Punizlan civilization on Dahl Radam, a wind-powered generating station. It was in full operation as a succession of masted sleds were released one at a time at one end of the station and rolled on a bed of disk bearings several kilometers long. In the ground was implanted a sequence of magnets stretching the entire length of the bed. Each sled bore two batteries connected to a wire loop which encircled the sled's base and passed through a circuit-breaker triggered by approaching magnets. As they proceeded along the bed, the batteries gradually became charged.

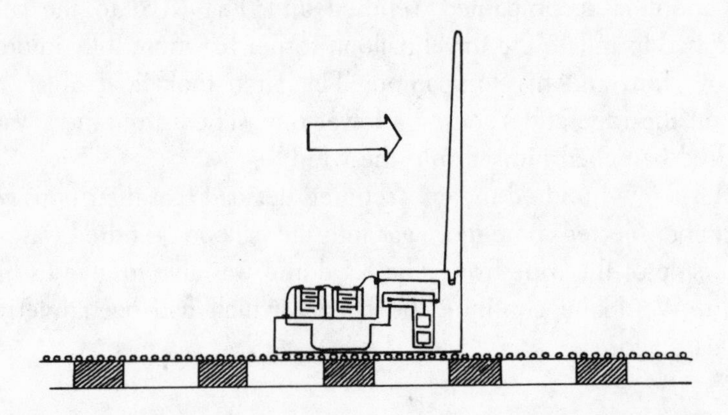

This charming operation was the recent result of one Punizlan's initiative. Why continue buying new batteries when the old ones could be recharged? Every day a freight balloon brought a fresh load of dead batteries and picked up the recharged ones.

Yendred and Tba Kryd had no choice but to use their balloon if they were to cross the power station while it was in operation. As they passed over the head of the technician whose job it was to release the sleds, he looked up at them and began to signal wildly with his arms. They were descending from their most recent hop right into the path of a power sled!

Yendred landed right in front of it and tried to hop away, but the top of the sled caught him just below his lower arm. Yendred had no choice but to hang on to the sled's sail, Tba Kryd and the travel balloon crowding down on top of him. With a mighty heave, Yendred pushed hard on the sail, launching the pair once more into the wind, where they would go slightly faster than the sled. For two more kilometers it was touch and go, with Yendred and Tba Kryd touching down and going up, each time barely avoiding the sled. It was like watching a chase scene in which some impassive mechanical monster bore down on two helpless humans.

When they had cleared the station, Yendred and Tba Kryd alighted for a rest.

▪ EAVESDROP

I IN A PASSENGER BALLOON TO BE WOULD LIKE [said Tba Kryd].

WE MANY MANY FSADS TO GO HAVE AND A PASSENGER BALLOON MOST CONVENIENT WOULD BE. BUT THIS WAY WE INDEPENDENT ARE AND MUCH WILL SEE. HOW YOU TO VANIZLA BEFORE DID GO?

BY PASSENGER BALLOON. I THIS COLD DO NOT LIKE AND I HUNGRY AM.

The two pulled out of their string bag a curious form of Punizlan bread that Alice called "jelly rolls," a kind of doughy strip baked into a spiral. Yendred and Tba Kryd both ate two. Then Tba Kryd ate one more as Yendred redeployed the travel balloon, which had been anchored in their lee. The afternoon wind was still building and Yendred was eager to be off.

Their next landmark was the jebb quarry, still more than a hundred kilometers distant. Operated by both Punizlans and Vanizlans, it formed their common boundary. Beyond that lay the ancient shrine of Amada and the vast stretch of Vanizlan Dahl Radam.

By now it was nearly dawn in the laboratory and we prepared to leave.

▪ WE MUST GO. BON VOYAGE.

WHAT IS ''BON VOYAGE''?

After the contact ended, Ffennell took a pad of paper and drew a row of eleven squares on it. Then he collected pennies and nickels from everyone in the room.

"The game of Alak," he said, "is basically just one-dimensional Go, a well-known board game here on Earth. Let's try a few games!" I was too tired to stay on, but Lambert and Chan remained to play. The other two followed me out of the laboratory. Dawn was streaking the early Friday sky.

10

Drabk the Sharak of Okbra

I did not get into my office on Friday until 11:30 in the morning. For what seemed like the hundredth time I thanked my lucky stars that our university had no summer term as such. Even its busiest department, computer science, offered only a handful of undergraduate courses, and our faculty were expected to use all their extra time doing research. In any event, we did not have to be continually visible.

Everyone, including my wife, thought (or hoped) that my weird hours and preoccupied air marked a deep commitment to some difficult and abstruse problem. It was touching how, as I left for work that morning, my wife had stood smiling a bit nervously by the door.

"Good luck, dear!"

When I got to my office, I found an envelope taped to the door. I opened it.

> Prof. Dewdney—
> Was here to see you.
> Back at noon.
> P. Craine

I hadn't seen Craine since our first contact with Yendred in the spring. One of the few students privy to the events of those days, Craine had been unable to stay over the summer. He had taken a job with a large computer firm on the West Coast at the end of May. I remembered Craine as a quiet, likable student who had carried an exceptionally heavy load of courses. I sat back in my swivel chair and fantasized for a few moments about how much fun it would be to bring him up to date on our contacts. This was his probable reason for dropping by.

A tentative knock on the door announced Craine's presence.

"Good afternoon, sir." He smiled awkwardly.

"Craine! Nice to see you again. How are things at. . . . watchama-callit?"

"Just fine, sir. I came East for a training seminar and thought I'd come by to see how everything was going."

"It's been amazing! I know you couldn't resist the lure of the almighty dollar, but you really should have stayed." I then lifted one eyebrow, as at a fellow conspirator. "By 'everything,' I assume you mean 2DWORLD."

"Yes."

He seemed a bit ill at ease, standing just inside the door. I showed him to a chair and carefully closed the door. Craine spent a minute studying the patterns in my carpet as I searched my desktop for a certain folder. He spoke just as I found it.

"Are you still in contact?"

"Yes, indeed! Here, take a look at some of these!"

I handed him the folder of Polaroid photographs. On these could be seen the faint outlines of some of the more exciting scenes we had witnessed. Craine took the folder on his lap, inserted a finger below the flap, but did not yet open it.

"I mean, are you still having regular contacts?"

"Well, pretty regular. I'd say two days out of three, on average."

"Is it still secret?"

"Yes, definitely. We had to really tighten up after some bad publicity. But, as far as I know, no one but our group—including you—really knows what's going on. Things have died down very nicely."

Craine opened the folder cautiously, glanced at the first photograph, and then closed the folder.

"How long do you think it will go on?"

"That's the big question!" I smiled, liking Craine in spite of his obvious shyness. "Maybe a bigger question is 'how long can *we* go on?' These late nights are hard on all of us, I'm afraid. We sit up all night talking to Yendred, and then we wander around bleary-eyed all day trying to act like normal university types. But go ahead and look at those pictures. The third one shows a kind of two-dimensional steam engine. Fascinating device!"

Craine did not open the folder.

"Say, are you all right?"

"Yes. I think so."

He raised his head slowly from the folder to look at me with an expression that I shall never forget. His eyes were a strange, opaque color, filled with a kind of silent pleading, a quality of fear and pain barely admitted.

"Don't you want to look at the pictures?"

"Not really." He paused and then took a breath. "They bother me."

"They what?"

He laughed with shocking suddenness. "You know, I really find it strange that you're continuing with this. It's all a fake, you know."

"Fake?"

He laughed again, blinking rapidly. "Yeah, fake. It's a joke."

"Well, whose joke is it? Who's the joker?"

"Don't you know?"

"No, I don't!"

Craine got up from the chair and laid the folder on my desk. "Yes, it's a joke. I'm afraid I have to go. I have to catch a plane. If I were you, I wouldn't do this too much longer or everyone is going to be laughing at you."

I did not argue with Craine because he seemed driven by some unknown force to say these things. Some inner compulsion was trying to laugh the Planiverse and its inhabitants away. He wore a peculiar glazed smile as he walked to the office door, turning to look at me and shake his head as he went.

"Maybe it *is* a joke." That was all I could think of saying.

Craine did not say goodbye but shook his head one more time and walked away down the corridor.

That look of opaque fear stayed with me. Something in it reminded me of the sudden gulf I had felt at the cottage, the split between the reality I shared with others and a new reality in which dreams walked and fantasies wore the cloak of truth. Between the two realities lurked the Planiverse. Where, then, was Craine's mind? Had it crossed the gulf to a fearful premise more acceptable than the Planiverse? Did it hurl laughter down into the abyss?

Later that afternoon, I met with the students for our weekly, but secret, "2DWORLD seminar." As soon as I brought up the subject of Craine, Lambert chimed in.

"I saw him on campus this afternoon. I yelled to him but he didn't hear me."

"What time was that?"

"About three."

Everyone wanted to know if Craine had wanted to visit the computer laboratory and renew his acquaintance with Yendred. I told them no, that Craine was flying out that very day. Our next contact would not take place until that evening.

Saturday, July 26, 9:00 P.M.

Shortly after the Earth/Arde link was established, the printer typed out nearly two pages of complaints from Yendred. He and Tba Kryd had been traveling all afternoon in a fierce wind with few stops for rest. Even now the two hung grimly from their balloon as it bobbled in the turbulent, late afternoon air. Their hops exceeded ten meters.

As sunset approached, the wind began to die away and, after a rather modest hop, they stumbled to a halt, Tba Kryd falling over in the process. Yendred almost let go of the balloon before he had a chance to deflate it.

```
THIS INSANITY IS. WE OF EXHAUSTION SHALL DIE. WHAT THE POINT OF
THIS TRAVEL IS? WE A PASSENGER BALLOON SHOULD HAVE TAKEN.

IF YOU A MEAL WILL MAKE I THE WINDBREAK WILL CONSTRUCT.
```

The two travelers no longer maintained the lofty dialogue of earlier days. By the time Yendred had constructed a windbreak to their east from the many angular stones and boulders which lay about, Tba Kryd had ignited a small "fire brick" and warmed some food. Ignoring good form while eating, the two discussed the quarry, now something like fifty kilometers to the east, and the best way of crossing it.

After supper, Tba Kryd excavated a shallow pit and ignited several fire bricks in the bottom. He and Yendred immediately refilled the pit with stones and laid the travel balloon and ropes over this. Presumably, the heat from the fire bricks would seep up through the rocks to warm their makeshift bed.

Before falling asleep they argued about how to cross the quarry.

By Sunday I was beginning to admit to myself that the Craine incident had bothered me more than I realized. I found myself being haunted by the old fear: if the Planiverse was not real, then I too had somehow crossed the gulf. Perhaps it was fortunate that what Yendred was about to encounter on the heights of Dahl Radam was so bizarre and terrifying: its very "otherness" reaffirmed the reality of Arde.

Sunday, July 27, 9:00 P.M.

When we finally established the 2DWORLD link, we saw Yendred and Tba Kryd standing on the surface, preparing to set out on another afternoon of travel. The ground on which they stood appeared to be a vast heap of boulders, below which we could see a kind of bedrock full of cracks and joints. The landscape had become craggy and irregular.

As the two perched on a boulder they argued again about how to cross the quarry. Tba Kryd wanted to descend by balloon to the quarry floor and to travel across it by the usual method; this would be slow going, however, because the morning calm was upon them. Yendred wanted to wait until just before noon and then inflate the balloon so fully that they would glide over the whole quarry.

HOW YOU AGAIN COME DOWN WILL?

I AIR FROM THE BALLOON WILL RELEASE. I THIS HAVE DONE BEFORE.

THAT MOST DANGEROUS IS. I TO SUCH A FOOLISH PLAN WILL NOT AGREE.

Tba Kryd won. As it happened, this was a bad decision but, at the time, we felt relieved that another aerial adventure was not in the offing.

They set out, much as on the previous contact, Yendred doing all the hopping and Tba Kryd hanging passively on the upper part of the rope. Perhaps Yendred did this out of deference to Tba Kryd's seniority. As we eavesdropped further on their conversation, it became apparent that Yendred was a little disenchanted with his companion.

HOW YOU YOUR PLAN FOR ARDE THE BEST IS DO KNOW?

IT OBVIOUS IS. WE ONE NSANA MUST BE.

WHY WITH TWO IT NOT BETTER IS?

Soon enough they grew quiet as the business of merely hanging on to the balloon ropes ocupied their attention. They traveled quietly for several hours until they at last reached the quarry. They stopped on the rim to survey the great pit before them. For centuries both Punizlans and Vanizlans had been cutting blocks of jebb from the quarry. In fact, the boundary between the two countries lay right in the quarry, rather closer to the Vanizlan cliff.

We scanned down and discovered the quarry floor not twenty meters below the pair. This was a modest descent by balloon. As a last-minute compromise, Yendred inflated the balloon just a little more fully so that their hops would at least be fewer and farther between.

Down they went, landing with a bit of a thump but rebounding directly into the air again. The floor of the quarry was littered with fragments of jebb and, here and there, a broken tool. They came to a narrow but rather deeper pit, dug to some forgotten purpose, but leapt over it easily. There was no wind, and their travel was correspondingly slower.

The thing that then entered the screen literally made our hair stand on end. Yendred and Tba Kryd had just begun another gentle eastward leap when an enormous snakelike creature, wriggling through the air, descended on their balloon. It was quite the largest living creature we had ever seen on Arde, apparently a gigantic Bes Sallur.

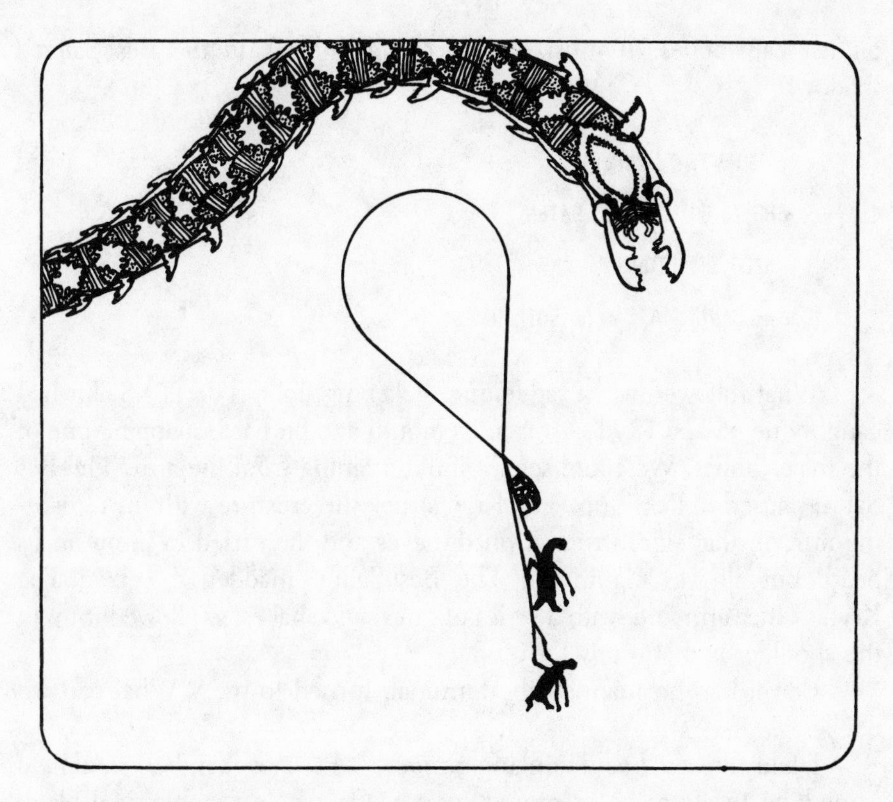

Its midsection landed upon the balloon, forcing it and the two hapless travelers to the ground. The snake extended its head and tail rapidly over the quarry floor, trapping everything beneath it.

BES SALLUR. IT'S A BES SALLUR. HELP US.

▪ CAN YOU FIGHT IT?

Yendred did not answer. He was struggling beneath the weight of the creature, which now began tucking both its head and tail beneath its body. It was already clear what it intended to do: having trapped its prey, the Bes Sallur had merely to eat them at its leisure. As chance would have it, Tba Kryd was now next to the Bes Sallur's mouth. The balloon, greatly squashed, had not yet broken. On the other side of the balloon was Yendred, unable to stand, but struggling against the creature and beating on its scaly bones. It must have been pitch-black in the Bes Sallur's shadow.

WHERE IS TBA KRYD?

▪ HE IS GOING TO BE EATEN.

STOP IT. STOP IT.

▪ WE CAN'T. WE CAN DO NOTHING.

What followed was a saddening, sickening sight. The Bes Sallur had caught one of Tba Kryd's arms in its mouth and bit hard, snapping one of the major bones. We could see the fluid draining from the arm. The Bes Sallur paused to lick it up. Tba Kryd struck the creature with his remaining arm on that side, struck it on the eyes and then tried to jump on its head. But all was nightmare. The Bes Sallur, maddened, seized Tba Kryd's other arm and, with a series of gulps and shakes, swallowed it up to the shoulder and abruptly bit it off.

Edwards, who manned the terminal, turned to me. "What can we do?"

I had already been thinking furiously. Perhaps Yendred could do something to close the creature's mouth. Or a pole. A pole or a plank would make a perfect shield.

"There's no pole about, is there?"

Tba Kryd lay motionless and the Bes Sallur ignored him for the time being, probing with his tail for Yendred on the other side of the squashed balloon. Alice had heard me. "If I had a pole, I'd shove it down his throat!"

Suddenly, Chan yelled. "Hey! That's it! The gas bottle!"

We all understood immediately.

▪ YNDRD. GET THE GAS BOTTLE.

WHAT IS THE BES SALLUR DOING?

▪ NEVER MIND. THE GAS BOTTLE IS UNDER THE BALLOON TO YOUR LEFT
[erase] EAST.

I had never seen anyone type so fast. Yendred leaned beneath the balloon, groping beneath the great snake's crushing weight. He had it! But now the Bes Sallur's pointed tail had found Yendred and the great head lunged forward, biting at the balloon. Suddenly the balloon was no more, just an air pocket in which Yendred faced great teeth in total darkness.

Yendred held the gas bottle in front of him.

▪ LET HIM HAVE THE BOTTLE. WHEN HE . . .

The Bes Sallur lunged, catching Yendred's hands and the gas bottle in his teeth.

"Oh, no!"

A scream of anguish from Alice as Yendred's lower hand came unjointed. He snatched his upper hand from the teeth as the Bes Sallur opened his jaws further to crush the morsel with its rear teeth. It bit down. Hard.

What happened next was most gratifying. With a great rush, the gas from the ruptured bottle inflated the creature's stomach, bursting it open. Fluid ran out its jaws and the great body writhed and quivered as random messages shot along shredded nerves and zipper junctions parted, one after another, down the length of the creature.

Yendred had barely the strength to crawl out from under the massive corpse. He did so by squirming west, emerging from beneath the still quivering tail. Tba Kryd was undoubtedly dead, his torn body lying next to the Bes Sallur's head. Yendred stepped upon the tail and walked shakily until he stood upon the summit of the creature's inert form.

▪ YOU BETTER GET YOUR STUFF. IT'S STILL UNDER THERE.

I WILL NOT GO UNDER AGAIN.

His lower left hand was torn and ragged. He pinched it in his upper hand to keep from bleeding and walked the remaining length of the creature, hopping to the ground at its eastern end. He was evidently in a state of shock. How would he survive without his food supply, at least?

THANK YOU MY EARTH FRIENDS. THANK YOU.

He continued to walk east along the quarry floor away from the creature, stumbling on the stones.

▪ YOU SHOULD REST FOR A WHILE.

I WILL NOT REST NEAR THAT BEAST. IT IS BETTER TO KEEP WALKING.

▪ ARE THERE MORE BES SALLURS AROUND?

Yendred stopped walking and carefully scanned the entire sky from horizon to horizon, all in one slow sweep. No. He saw none. We decided to look for some by backing off the scale so that Yendred dwindled almost to a dot and we could see much of the quarry. He continued to walk east as we anxiously scanned the screen for more of the flying predators.

"There, look!"

Curled up in one of the pits which Yendred and Tba Kryd had ballooned over was a motionless Bes Sallur. It had apparently failed to waken during the pair's quiet overhead passage. Had the one which attacked also been sleeping in a pit?

▪ THERE IS ANOTHER BUT IT IS PERHAPS ASLEEP. IT IS LYING IN ONE OF THE PITS YOU CROSSED. AHEAD OF YOU THERE ARE NONE. DO YOU SEE THE QUARRY WALL AHEAD OF YOU?

We came in close to Yendred again. He continued to hold his hand and his steps seemed shakier than ever. There seemed to be no way he would ever scale the eastern wall of the quarry. When he at last arrived, he appraised its smooth, impassive surface from his vantage point and we examined it from ours. There were no handholds, and Yendred, with his injury, was in no shape to climb it even if there had been. Yendred had no food or shelter and the sleeping Bes Sallur might awaken at any time. I felt a curious hollow in my stomach, as if this might be the end of everything. We would come into the laboratory on the next few nights either to find a gradually weakening Yendred or to be entertained only by the antics of the denizens of 2DWORLD. Finally, sooner or later, the Earth/Arde link would dissolve forever.

We stayed with Yendred until Shems had set on the quarry and the bitter cold of Dahl Radam's night began. We stayed even a few hours beyond this, when dawn began to lighten the eastern sky outside our window. We talked with Yendred about anything and everything, trying to keep his spirits up.

▪ WHERE WILL YOU GO FROM THE QUARRY?

THE SHRINE. IT IS JUST A FEW MORE DAYS' JOURNEY.

▪ DO YOU HOPE TO MEET DRABK THERE? TELL US ABOUT HIM.

DRABK LIVES ALONE SOMETIMES NEAR THE SHRINE AND SOMETIMES HE IS SEEN IN THE CITY OF OKBRA. HE HAS THE KNOWLEDGE OF BEYOND AND THE KNOWLEDGE BEYOND THOUGHT. HE IS KNOWN TO HAVE WORKED MIRACLES.

▪ WHAT MIRACLES?

HE CAN DISAPPEAR IN ONE PLACE AND APPEAR IN ANOTHER. HE CAN HEAL SICK NSANA. HE KNOWS THE PAST AND THE FUTURE AND HE CAN FLY.

▪ HOW DO YOU KNOW THAT HE CAN DO THESE THINGS?

I HAVE READ OF THEM AND TWO YEARS AGO I MET ONE WHO SAW DRABK DISAPPEAR.

▪ PERHAPS DRABK IS A MAGICIAN.

DO YOU MEAN HE DOES NOT MAKE REAL MIRACLES?

At this point, I tapped Ffennell's shoulder. "That's enough. Stay positive."

Soon enough it was time to go. Alice, Edwards, and Chan wanted to stay on and I had to become quite stern on this point.

"It's true Yendred's life is in danger, but then again, he might survive the ordeal. But if you want this whole operation closed down for good, then stay. By all means, stay!" I needed sleep.

Home once again, I went to bed but slept poorly. There was a recurring dream in which I would start up the 2DWORLD program and, instead of Yendred, Craine would appear on the screen in a sort of gruesome cross section, struggling to be free of his stifling planar world.

Monday, July 28, 9:00 P.M.

We arrived in the computer laboratory all at the same time and quickly filed in. Tension filled the air. None of us knew what condition we would find Yendred in. If he was already dead, whether from exposure or from the attack of the Bes Sallur, we would never reach him at all.

A portion of 2DWORLD's linear landscape reassembled itself into the craggy quarry bottom and there sat Yendred, still alive! All six of us cheered spontaneously. Soon Shems would be rising over the quarry and he would have made it through the night.

▪ HELLO. HELLO. HELLO. ARE WE GLAD TO SEE YOU.

WHY DO YOU ASK?

▪ NO. I MEAN WE ARE REALLY GLAD TO SEE YOU ARE STILL ALIVE. HOW IS YOUR HAND?

YOU CAN UNDOUBTEDLY SEE FOR YOURSELVES THAT IT HAS STARTED TO HEAL. AT LEAST THE BODY GLUE IS HOLDING. BUT I CANNOT SPEND ANOTHER NIGHT LIKE THIS.

Just as Yendred said this, there came a knock on the laboratory door. We froze.

I watched from the terminal as Lambert went to the laboratory door and looked through the glass. "It's the chairman!"

▪ YNDRD. WE JUST WANTED TO CHECK WITH YOU. KEEP YOUR COURAGE. GOT TO GO.

I went to the door. Chan took my place at the terminal.

"May I come in?"

"Certainly. Of course. I, ah, was working late this evening and Alice Little asked my opinion about her data base management project. It's really quite interesting!"

The students shrank away from the chairman's presence as he strode around the corner of the memory cabinet to confront the terminal and display screen. There sat Chan maneuvering the thrusters of the spaceship Enterprise. He launched three successive photon torpedoes at the nearest Klingon battle cruiser, and looked up with what he undoubtedly hoped was vacuous innocence.

"Miss Little, is this your data base project?"

Alice did herself credit with a very genuine-sounding laugh. "No, sir. Chan and Lambert wanted to play Star Trek while I talked to Professor Dewdney about my project. I'll have to kick them off the computer soon."

"You'll have to kick them off right now, I'm afraid. This is a research computer, not a toy." His tone softened. "Look, you all know department policy about computer games. I had hoped never to see this sort of thing again." He turned to me. "Didn't you see what was going on?"

I smiled ingratiatingly, rather glad underneath that the real "game" was over for now. "I'm afraid I did. I guess I didn't feel there was any harm in it. These are all hard-working students, and Lambert and Chan seemed in the mood for a break."

"As far as I know, Little is the only one with any business in this room."

By now, Chan had killed the program. He and Lambert left the room. Ffennell and Edwards stood by the door, listening.

"I have received word that something funny is going on in this laboratory in the evenings. Once again, Dewdney, I find you very much in the center of things. Miss Little, how *is* your data base project going?"

Alice rattled on about her work at some length. Presently the chairman began to look slightly bored.

"Very well. Continue with your discussion. I'm going to my office to work on a report." He gave me a strange, sidelong glance as he left, as though no longer sure that he hadn't stumbled onto a scene as innocent as it appeared. Perhaps the "word" he had received had been vague enough to include computer games.

The chairman's visit and continued presence put an end to our contact that evening. As soon as he left the room, Alice muttered, "At least Yendred is still alive!"

On the next evening, we decided to start our contact period much later than usual.

Tuesday, July 29, 11:45 P.M.

We kept trying to establish contact but did not succeed for several hours. By our calculations, the normal Ardean sleep period would end sometime after 3:00 A.M. by our laboratory clock. Still, how could Yendred sleep in the cold of the quarry at night? It became increasingly likely that he had not survived. Every fifteen minutes, Lambert would start up the 2DWORLD program and wait for a response.

▪ HELLO YNDRD.

UNKNOWN: ''HELLO.''

Then, at 3:49 A.M. came the reply as Yendred appeared. He was in a small underground house lying on a bed. Nearby sat an old Ardean who looked at him from time to time.

HERE I AM. I WAS RESCUED.

▪ HEY, YOU TWO-DIMENSIONAL FREAK. YOU REALLY HAD US WORRIED!

I allowed Lambert his dubious pleasure. Yendred's narration was more important. It turned out that a group of Vanizlan quarry workers had arrived shortly after we broke off contact on the previous Arde morning. They had found a very cold and tired Punizlan crouching at the bottom of the cliff. Taking him back to the hut of an old Vanizlan who lived nearby, they had warmed and fed him, then reset his hand and dressed it. He guessed that he had slept most of the previous day as well as through this latest night.

The old Vanizlan handed Yendred a bowl containing a sort of stew. It was some time before Yendred would talk to us, so absorbed was he in the food. Then he spoke as the food digested in his stomach.

THESE ARE THE MOST KIND NSANA. IT IS VERY PRIMITIVE HERE BUT I
BEGIN TO FEEL AT HOME. THIS OLD ONE IS MOST PLEASANT.

He went on at some length about the old one's kindness to him and how he would be content to live with the quarry workers forever.

▪ WHAT ABOUT GOING TO SEE DRABK? WHAT ARE YOUR PLANS?

OF COURSE. I WILL LEAVE AS SOON AS I AM WELL.

Presently, when all the food had been processed in his stomach, he spat up the remains into a pot near the foot of the bed. He stepped over it and raised his arms, taking a deep breath, so to speak. Now, by Ardean protocol, it was permissible to speak.

▪ EAVESDROP

I MUCH BETTER FEEL.

YOU HERE LEAVE NOT SHOULD UNTIL YOU BETTER THAN BETTER FEEL. YOU WHERE DO GO?

I TO THE SHRINE DO GO.

A PUNIZLAN THE RELIGION OF AMADA FOLLOWING. THAT MOST STRANGE IS.

MY REASON RELIGIOUS NOT IS. [Here there was a long pause.] I DRABK SEEK.

AH. DRABK THE SHARAK OF OKBRA. HIM YOU SEEK AND SEEK AND SEEK MAY.

BUT HE NEAR THE SHRINE LIVES.

THAT TRUE IS EVEN WHEN HIM YOU DO NOT SEE.

HE INVISIBLE IS?

IN A WAY OF TALKING YOU THAT MAY SAY.

The old Vanizlan told Yendred that his hut was a day's walk from the shrine. He instructed Yendred in the protocol to be used by a foreigner in visiting the shrine. There was a block of jebb set into the ground near the shrine. Beyond this, Yendred was not to go unless invited by a Vanizlan. Moreover, he must under no circumstances attempt to cross the shrine alone even if he was foolish enough to pass the stone marker. Apparently, something truly terrible would happen if he tried.

It was rather annoying that Yendred then decided to go to sleep, wasting the rest of our session. Before he dozed off and contact was broken, we recorded the Vanizlan's hut and everything in it. There were only two floors, and the supporting beams appeared to be made of hatr. There were two beds, a stair that swung but had no spring, a table with a simple rock-bed cooking stove on it, a shelf with some bowls, and another shelf with five books piled up on it. There was a single electric light on the table, plugged directly into a battery which acted as a stand. Except for the electric light, I imagined that we were looking at the sort of rude house in which all Ardeans had once lived.

But then contact was broken as Yendred fell asleep.

We cancelled the next evening's session because Yendred would also be asleep for much of that time.

Thursday, July 31, 9:00 P.M.

Unknown to us, our contact with Yendred and Arde was about to take on a new dimension. By now we had become adjusted not only to the bizarre world of Arde but to our almost daily interaction with it, and all of us in our heart of hearts wondered if there were any further major surprises in store. There were.

When we established the Earth/Arde link shortly after 9:00 P.M., it was late morning on Arde and we found Yendred standing upon a buried stone looking toward the shrine and speaking. There were no other Ardeans around. At Yendred's feet lay a small string bag with some food in it.

▪ WE ARE BACK. WHO ARE YOU TALKING TO?

I'M CALLING THE NAME OF DRABK. DO YOU SEE HIM ANYWHERE?

We scanned east to the shrine and beyond, then west past Yendred. No other Ardeans were to be seen. Yendred was disappointed, but continued to call out every few minutes.

The shrine was the only true building on all of Arde—at least in the earthly sense of the word "building." It was made of stone blocks and stood well above the surface.

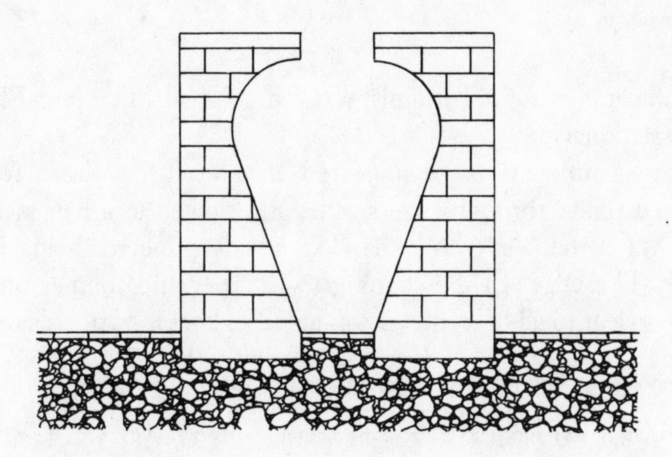

It was certainly in the logical spot for Arde's only building. Whatever rain fell here on Dahl Radam's heights was shed away from the shrine. In times past, before the development of travel by balloon and rocket, it had stood as a barrier between the Punizlans and Vanizlans. None could cross the shrine unless they adopted the Vanizlan religion and became, in effect, Vanizlans. Given the current Punizlan view of things, very few were crossing the shrine these days—at least not by land.

The shrine had an entranceway at its top, but was empty inside. What function did it serve besides separating the two nations of Arde?

Suddenly Alice spoke. Her voice sounded both puzzled and ominous.

"Damn it. There it is again. I keep thinking it was a tube malfunction. Did you see that shadow?"

We hadn't. Yendred continued to call every few minutes between times speaking with us.

▪ HOW LONG WILL YOU GO ON CALLING?

I WILL CALL ALL DAY IF NECESSARY.

Then we all saw the shadow, a vague, fuzzy patch gliding slowly past Yendred, then past the shrine and off the screen to our right.

▪ DID YOU SEE ANYTHING GO BY- LIKE A CLOUD?

NO. DID YOU?

Whatever it was, it certainly was no tube malfunction. Phosphor burns are stationary.

The next time the shape appeared, it entered the screen from our right, passed right "through" the shrine, and slowed to a halt in front of Yendred. What followed was rather like a slide projector being brought into focus. The edges of the shape grew sharper and sharper until suddenly it was clear to all of us that it was about to become an Ardean shape.

▪ DO YOU SEE ANYTHING IN FRONT OF YOU?

NO. BUT YOU ARE MAKING ME VERY NERVOUS. I FEEL VERY STRANGE. LIKE SOMETHING IS GOING TO HAPPEN.

With a kind of snap, the Ardean image lost its remaining fuzziness and became sharply etched on the screen.

"Drabk the Sharak of Okbra," murmured Alice.

▪ EAVESDROP

YOU THIS ONE'S NAME HAVE CALLED.
[Yendred took almost two minutes to reply.]

I HONORED AM THAT YOU CAME.

YOU FAR HAVE TRAVELED AND NOT ALONE.

I ONE WITH ME HAD. TBA KRYD. BUT HE DEAD IS.

THIS ONE THAT ONE DOES NOT MEAN. THERE OTHERS ARE THERE ARE NOT?

[Again, Yendred paused a long time.]

YOU THE ONES OF EARTH MEAN?

YES.

Here was an astounding turn of events. Quite apart from Drabk's magical entrance, arranged by what artifice I could not then imagine, he seemed to know about us. Had Yendred already secretly met Drabk and told him about us?

> THEY WITHIN THE CIRCLE OF CIRCLES DWELL. IN THE SPACE BEYOND SPACE. I THEM THE KNOWLEDGE OF BEYOND ME TO TEACH HAVE ASKED BUT THEY OF THIS NOT TO KNOW SEEM.
>
> TRULY THEY LESS THAN YOU KNOW. [Pause.] YOU WITH A PURPOSE HERE CAME?

We all sat before the display screen, six mesmerized beings, eyes glancing back and forth between the screen and the printer.

> I THE KNOWLEDGE OF BEYOND SEEK. WHAT THE KNOWLEDGE BEYOND THOUGHT IS?
>
> THIS ONE ANSWER CANNOT. THE KNOWLEDGE BEYOND THOUGHT BEYOND WORDS ALSO IS. BUT KNOWLEDGE OF THE BEYOND THE KNOWLEDGE BEYOND THOUGHT FOLLOWS. AND ALL BY PERMISSION OF THE PRESENCE IS.
>
> WHERE THE PRESENCE IS?
>
> IT BESIDE YOU IS. CLOSER THAN YOUR BODY. CLOSER THAN YOUR BLOOD. THIS ONE TO THE EARTH SPIRITS CAN POINT BUT NOT TO THE PRESENCE.
>
> YOUR ARM GONE IS.

Indeed, one of Drabk's arms had become very fuzzy and had then shortened to nothing. Presumably, it had become invisible to Yendred.

"You know what I think?" said Chan. "That guy can move into the third dimension!"

Chan's remark made a strange sort of sense, although I must say that for some reason I felt reluctant to admit the possibility. But from a purely naïve interpretation of the visual image before us, it did appear as though Drabk's arm was swinging out into a third dimension to point directly to us!

Abruptly Drabk's arm swung back to its original position.

FEW NSANA SUCH THINGS HAVE SEEN AND DANGEROUS IT IS TO SHOW.

WHY THAT IS?

A MIRACLE NOTHING PROVES. IT THE IMAGINATION STRONGER MAKES
BUT THE UNDERSTANDING WEAKER.

WHY YOU ME THIS THING DID SHOW?

YOU FOR THE KNOWLEDGE APPOINTED ARE AND LITTLE TIME REMAINS.

On asking how Drabk knew that he, Yendred, was "appointed for knowledge," Yendred learned that Drabk had "seen" his future and already knew what would come to pass. Beyond this, there was a long discourse by Drabk on "wooing the Presence." This sounded almost like a courtship ritual, albeit with some unseen, all-powerful entity.

Drabk further explained that it would soon be sunset and that Yendred must decide before that time whether he would seriously seek the "knowledge beyond thought." Drabk did not reveal precisely how long the process would take in Yendred's case, although he did hint at a matter of weeks for a process which normally required years for its completion. Here was a completely new twist in Yendred's quest. Although very much a skeptic about what some might consider to be parallel belief systems on Earth, I had to admit to being fascinated by Drabk's teaching. Although I would not have been surprised to learn that the old Vanizlan was indulging in some sort of fakery, I found his statements strangely compelling. I suppose we all did. There, after all, to my left was Lambert the video-game hotshot staring slack-jawed at the screen.

There was a long silence from the printer as Yendred walked slowly away from Drabk and back to the stone marker. He stood upon the stone for several minutes with his head pointing up. Then he swung it to the west, as though looking for Punizla on the distant horizon. He came back.

I THE KNOWLEDGE MUST SEEK. THIS YEARNING NO OTHER OBJECT HAS.

Drabk raised all his arms simultaneously, as though taking a big breath. Slowly they returned to his sides.

WE THEN BEGIN. FOOD FROM YOUR BAG TAKE AND EAT. YOU THE NIGHT
IN THE SHRINE MUST SPEND.

All was silence again as Yendred ate from his string bag near the marker. Drabk stood immobile beside the shrine. By our calculation it would be sunset at the shrine in about twenty minutes. In the laboratory it was 2:35 A.M., the morning of August 1. We could hear a faint whirr from the electric wall clock blending with the hum from the computer.

"I have a feeling," whispered Alice to no one in particular, "that Yendred will not be with us for much longer."

"How come?" Another whisper.

"I don't know, exactly. I sometimes get the feeling that we've been used. You know what I mean?"

"Nope."

When Yendred had finished eating, he sat digesting for ten minutes and then carefully spat up on the other side of the marker. When he returned to Drabk, the latter reached out and took Yendred's upper hand in his own.

AFTER ME REPEAT. THERE NO PRESENCE IS BUT THE PRESENCE. THERE
NO KNOWLEDGE IS BUT THE KNOWLEDGE.

Yendred dutifully repeated these things, and then Drabk began to instruct him about what he would do in the shrine during his vigil there. Unfortunately, we shall never know exactly what Drabk said of this occasion as some kind of interference arose on the printer. For several minutes it jammed on the right-hand side of the paper, printing over and over again on the same spot, as though unable to process a carriage return or line feed. Panicking at the thought of losing contact at this crucial point, we were about to bring down 2DWORLD and put up a diagnostic program when the glitch disappeared and we caught the last of Drabk's instructions.

AND THE PRESENCE YOUR BEING PERVADES. WHEN THE PRESENCE TO YOU
MADE KNOWN IS THEN IN THAT RADIANCE WITHOUT THOUGHT YOU PERHAPS
THE FIRST KNOWLEDGE WILL RECEIVE.

Privately, I had been wondering how Yendred would enter the shrine. The walls were too smooth to climb and no ladder lay about. It was then that Alice gasped as Drabk and Yendred, still holding hands, rose into the quiet of the early evening air.

"Oh, God, this is a fairy tale!"

> FLYING IN PRINCIPLE SIMPLE IS. ONE MERELY ONESELF BEHIND LEAVES. FLYING THE FIRST ART IS— AND THE LAST REFUGE.

They floated up over the shrine and descended gently through the opening to the floor. Drabk gave a last instruction to Yendred that what he should at first blindly repeat he should strive continually to understand. He should at all costs avoid falling asleep and to concentrate on the words. This was, said Drabk, the most auspicious spot on Arde for this exercise. Then Drabk left, not by flying out of the shrine and not by becoming a fuzzy patch, but by simply vanishing. Instantly.

"When is Yendred's next sunrise?"

Ffennell consulted a chart and punched his calculator. "Eighteen forty. That's six forty this evening."

It was now a little after four in the morning. Yendred sat motionless in the shrine while we worried about the next contact. The machine would be booked by others until eight o'clock that evening. We called Yendred to inform him of our next contact.

> ▪ HELLO YNDRD.
>
> PLEASE DO NOT DISTURB OUR FRIEND.

11

Higher Dimensions

We all met in an unused lecture room on the afternoon of August 1 for our regular Friday 2DWORLD seminar. Besides discussing events of the past week or planning contacts for the next one, we would sometimes examine a particular Punizlan device or probe a phenomenon of two-dimensional science. We would also sometimes talk about the Ardeans themselves, their social patterns and individual behavior. It therefore was not surprising when, on this particular occasion, Edwards took the floor to outline his "theory of dimensions."

Edwards had dark, luminous eyes which always impressed me as masking, or rather illustrating, a special sort of intelligence. There were circles under them on this day.

"I've been, ah, thinking about dimensions in connection with what happened last night. The funny thing is, I've had to start thinking in four dimensions and even higher ever since this train of thought began."

Someone remarked that his train looked as though it had been running all night. Edwards smiled weakly and proceeded to draw a perspective view of the shrine on the blackboard. It was fairly clear that he meant us to be standing just outside the Planiverse, as it were, and looking along this particular portion of the cosmic balloon's skin.

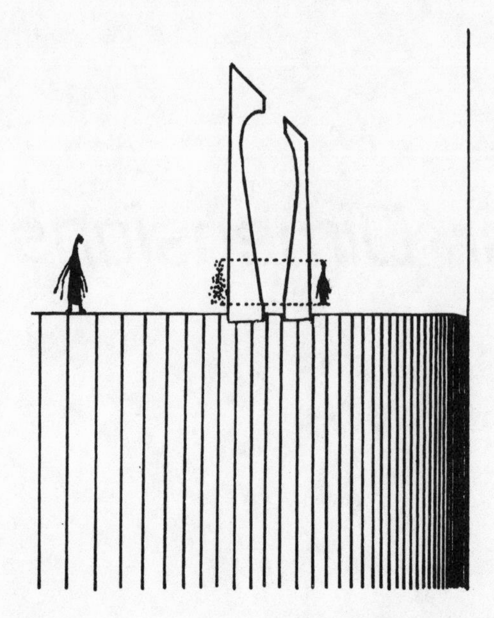

Edwards appeared to be imposing a scheme of dimensions in a very literal sense. Drabk, he said, had the ability to leave the Planiverse entirely and to glide along beside it. Even his distance from the Planiversal skin could be measured by the degree of fuzziness of his "shadow," as it were: like a hand in front of a projector, the closer Drabk came to the Planiverse, the more sharply defined his image became on our screen. Through all of this Chan was nodding his head in agreement. "That must be our space, too."

"Well, maybe not. Even if our own universe and Drabk's space have the same dimension, it doesn't mean there's any physical connection be-

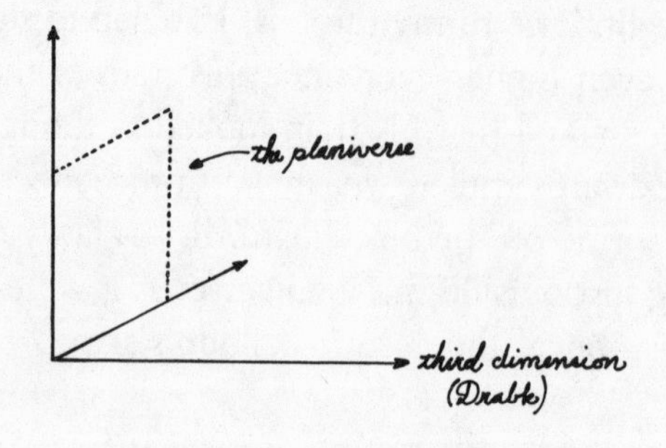

tween them. They could even be parallel spaces, separated by a fourth dimension. But that's where the fun begins."

Edwards next drew a set of axes on the board.

Two of the axes represented the two dimensions of the Planiverse. A third axis, intended to be at right angles to the other two, pointed in the direction of the third dimension through which Drabk habitually flew.

"So how did he disappear last night?"

"I'm coming to that."

Edwards drew a fourth axis. This was a fourth dimension, inaccessible to our view, into which Drabk had disappeared.

"Hey, maybe that was the time dimension."

"Maybe. Maybe not. But if Drabk has access to *all* dimensions, and if time is also a dimension, then he would have no problem going forward into the future or backward into the past."

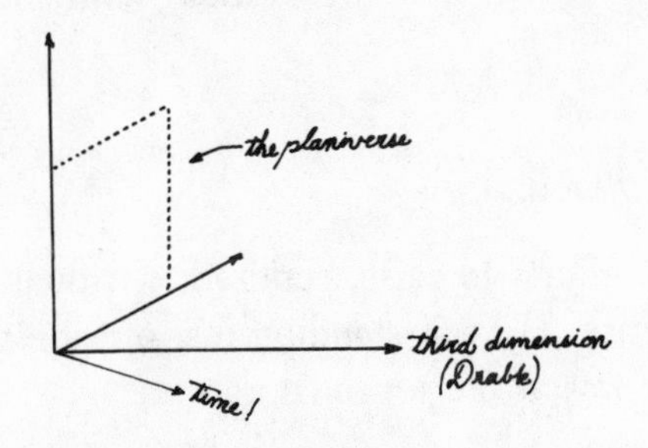

"You expect us to visualize that?"

No, Edwards didn't, because human beings were equipped to think only in three dimensions, just as Ardeans could only understand two. From the point of view of each dimension, though, all of the "lower" ones, no matter how many there were, would comprise a space which appeared thin, insubstantial, and, in a certain sense, nonexistent and unreal.

"I mean, just ask yourself this: if the Planiverse went right through the middle of this room, would anybody see it? No. It has zero thickness. It simply would not interact physically with anything in this room. Atoms, photons of light, whatever! From our point of view, it just wouldn't exist."

Here Edwards paused and looked at the ceiling and smiled. "Well, I guess it would exist, but it would be totally undetectable by us."

"So what about the big Presence that Drabk was talking about? Is that in some other dimension?"

"Well, it *does* fit in, but in a funny kind of way. Whether the total number of dimensions is infinite or not, one of these might be a sort of privileged dimension inhabited by . . . something or other which somehow manipulates what's going on in all the others. From the point of view of that something, its own reality would predominate and would be in intimate contact with all the realities of the lower dimensions—or other dimensions, anyway. All the manifestations of this one reality would inhabit a kind of multidimensional skin, including the time dimension, a really incredible, frozen pattern of existing things—including us, I suppose."

At this point I had to interrupt in order to remind the students that we had some planning to do. We had to plan for a session that very evening and one for Saturday evening. Sunday coincided with Yendred's sleep period.

Friday, August 1, 8:00 P.M.

After continued efforts to raise Yendred, we finally succeeded well after the Ardean sunrise. He was standing just outside the shrine, Drabk beside him. Lambert was at the terminal.

▪ WE ARE BACK. WHAT HAPPENED?

IN THE SHRINE I HAD A STRANGE EXPERIENCE THAT I CANNOT TALK ABOUT.

▪ WHY CAN'T YOU TALK ABOUT IT?

DRABK SAID IT WAS FOR ME ONLY TO KNOW ABOUT. HE SAID TO TELL NO ONE.

▪ DO YOU DO EVERYTHING DRABK TELLS YOU?

OF COURSE. HOW ELSE WILL I LEARN?

▪ WHAT ARE YOUR PLANS FOR TODAY?

I AM TO LEARN ABOUT FLYING.

▪ HEY. THAT SOUNDS LIKE FUN. CAN WE WATCH?

EXCUSE ME. [Long pause.] DRABK SAYS YOU ARE TO GO AWAY PLEASE.

▪ WE WANT TO STAY.

"Don't be silly," I cautioned. "Ask him why we must go way."

I AM SORRY THAT YOU WANT TO STAY.

▪ WHY MUST WE GO AWAY?

IT IS A PRIVATE THING. IT WOULD BE BAD BOTH FOR YOU AND FOR ME IF YOU WATCHED.

▪ WHY IS THAT?

UNKNOWN: ''WHY IS THAT.''

"That didn't take long," sighed Alice.

There was no more contact that evening. We tried every half hour until after eleven o'clock. We left the laboratory after the last attempt, Edwards falling into step with me as I left the building for the parking lot.

"What did you think of my dimension theory?"

"Well, I have to admit it's an intriguing possibility. I am tempted to lecture you like a scientist and urge you to think of it as a 'working hypothesis,' but we're very far from a classical experimental setup. All we can do is observe, and maybe not much longer."

"Do you think it's all going to be over soon?"

"I've had that feeling, but then the optimist in me comes out. This may be just another adventure of Yendred's. I must say that I was, and still am, looking forward to his traveling deeper into Vanizla. If this Drabk is any sample of what to expect . . ."

"Do you think that Drabk has unusual powers?"

"To be perfectly frank, I don't know what to think. I certainly resist the notion that Drabk can in some way disobey the laws of Planiversal physics, whatever they are."

"I've been reading about certain people right here on Earth who sounded a bit like Drabk. They—"

"Look, Edwards, Arde is Arde and Earth is Earth. Don't get the two mixed up or you may get very mixed up yourself."

I said goodnight and hurried on to my car. The night sky was dark and the moon luminous.

Saturday, August 2, 9:00 P.M.

We tried to get Yendred for two hours and then gave up.

UNKNOWN: ''YNDRD.''

Sunday, August 3, 9:00 P.M.

The students were depressed. I was depressed. What a way for it all to end! How many more evenings would we try before we finally gave up? Out of desperation, we decided to try on this evening also, even though Yendred would be asleep unless we contacted him fairly soon.

He wobbled into view, sitting once more in the middle of the shrine.

▪ HELLO. WE ARE BACK.

Yendred stood up.

HELP ME. SOMETHING TERRIBLE HAS HAPPENED.

▪ WHAT HAS HAPPENED?

IT IS FRIGHTENING BEYOND IMAGINATION. IT TERRIFYING IS.

▪ WHAT IS TERRIFYING? WHERE IS DRABK?

HE GONE IS AND THINGS THERE ARE WHICH I DO NOT WISH TO KNOW.

▪ WHAT THINGS?

I CANNOT SAY. EVEN TO SPEAK ABOUT THEM MAKES ME DOUBT EVERY-THING.

▪ WHAT THINGS?

THE PRESENCE IS NOT WHAT I THOUGHT IT WAS. IT REAL IS. IT REAL IS. IT REAL REAL REAL REAL REAL REAL↑C

[Fearing a malfunction, we deliberately killed 2DWORLD
in order to restart it.]

▪ READY.

RUN 2DWORLD.

▪ YNDRD. ARE YOU THERE?

DON'T LEAVE ME PLEASE. DON'T EVER LEAVE.

Yendred had obviously been confined to the shrine again. He reminded me of a lonely cultist locked away in a shack. There was no way he could climb out of the shrine, and he had obviously not yet learned to fly.

▪ CAN YOU FLY?

NO. I DON'T WANT TO FLY BECAUSE I WOULD HAVE TO USE THE
KNOWLEDGE.

▪ WHAT KNOWLEDGE?

WHO IS THERE NOW? IS EVERYONE THERE MY EARTH FRIENDS? IS ALICE
THERE?

▪ HELLO YNDRD. IT IS ALICE.

We talked thus with Yendred for hours. What was Drabk up to? Yendred was terrified out of his mind, terrified by some mental state that he had not bargained for. But we could not rescue him from it. We could only stay with him until the Ardean dawn.

"What time is Yendred's next dawn?"

"Uh, that would be nine twenty-nine."

"But it's Monday morning. Everyone will be coming in to work!"

"Damn."

Our own dawn now rose like a glorious doom in the eastern sky. Alice looked up from the keyboard. "Maybe we could risk it just this once."

"Stay around until nine thirty?"

"What do you think?"

▪ WE WILL STAY WITH YOU UNTIL DAWN.

"What if Drabk *never* comes back?" It was Lambert.
"Don't be stupid," said Alice.

DRABK SAYS THAT SOON I WILL TALK WITH YOU NO MORE. HE SAYS THAT WHEN I HAVE REACHED A CERTAIN STAGE I WILL NOT WANT TO TALK WITH YOU. I WILL HAVE SERENITY AND KNOWLEDGE BUT YOU ARE CLOSER TO ME EVEN THAN MY MOTHER AND FATHER. YOU PROTECT ME AND HELP ME. YOU ARE MY COMPANIONS. JUST TALKING TO YOU I DO NOT HAVE TO THINK ABOUT THE KNOWLEDGE. IT LIES IN WAIT LIKE A RA NIFID. IT SWOOPS DOWN LIKE A BES SALLUR WHEN I DON'T EXPECT IT. I WILL BE DEVOURED BY THE KNOWLEDGE AND SPAT UP AGAIN ALL CHEWED.

During this phase of his vigil, Yendred paced back and forth across the narrow floor of the shrine, talking incessantly. He seemed for all the world like a caged animal being stalked by some unseen hunter. Abruptly he sat down. It was nearly dawn at the shrine.

AND YET I WANT THE KNOWLEDGE. HOW IS THAT POSSIBLE? IT IS SO FRIGHTENING THAT MY MIND BECOMES ONE DEEP PIT OF FEAR. YET WITHIN THE KNOWLEDGE THERE IS A CERTAIN GLORY.

He stopped talking. What should we do? Already the secretaries would be unlocking the office door. Maybe he would be all right now and we could leave.

▪ YNDRD ARE YOU ALL RIGHT NOW?
YNDRD ARE YOU ALL RIGHT NOW?
YNDRD ARE YOU ALL RIGHT?

There was no answer.

By our calculation, it was almost exactly when dawn began to break on the heights of Dahl Radam that Drabk suddenly appeared beside the shrine. Simultaneously a student walked into the laboratory.

YOU NOW MAY EMERGE.

"What's going on? I'm booked on this machine for nine thirty."

"We have a special project going. Can you come back later?"

"Hey, what's what? Weird graphics you guys have. Hey, is this 2DWORLD?"

The students were alarmed. Yendred began to rise within the shrine. The intruder's jaw dropped.

"Let him watch," I said. "Let everyone watch. Call the chairman!"

Alice looked at me and smiled. She understood, too. Yendred descended to the ground beside Drabk.

I TO THE KNOWLEDGE HAVE SUBMITTED.

SO YOU HAVE. THIS THE FIRST STAGE COMPLETES.

WHAT BEYOND THIS THERE COULD BE? THIS ONE WHERE FEAR AND JOY ONE ARE HAS BEEN. ALL THIS NOTHING IS.

[Yendred looked around him vertically.]

THEN WE TO SOMETHING SHALL TRAVEL. TO YOUR EARTH SPIRITS FAREWELL SAY.

GOODBYE EARTH FRIENDS.

▪ WHEN SHALL WE TALK AGAIN?

She knew what was coming, but at least she tried.

WE CANNOT TALK AGAIN. TO TALK AGAIN IS OF NO BENEFIT.

▪ BUT WE HAVE SO MUCH MORE TO LEARN FROM YOU.

YOU CANNOT LEARN FROM ME. NOR I FROM YOU. YOU DO NOT HAVE THE KNOWLEDGE.

▪ WHAT KNOWLEDGE?

The laboratory was becoming crowded with students. One of the secretaries who had brought me a letter to sign remained to stare at the screen. Two of my colleagues, having heard of something going on in the computer laboratory, entered the room.

THE KNOWLEDGE BEYOND THOUGHT OF THE REALITY BEYOND REALITY.

▪ WOULD IT HELP IF WE LEARNED YOUR PHILOSOPHY AND RELIGION?

IT HAS NOT TO DO WITH WHAT YOU CALL PHILOSOPHY OR RELIGION. IF YOU FOLLOW ONLY THOUGHT YOU WILL NEVER DISCOVER THE SURPRISE WHICH LIES BEYOND THOUGHT.

▪ WHAT SURPRISE?

Alice was trying to draw Yendred out, to wring a few more moments of conversation from the luminous being on our dark screen. But already he and Drabk had become fuzzy. Everyone in the room crowded toward the terminal to watch as two vague patches floated off the screen. Then the shrine wobbled and dissolved. The irregular ground of Dahl Radam was replaced by a straight, horizontal line.

UNKNOWN: ''WHAT SURPRISE.''

A lone FEC hobbled mechanically into view. Alice got up from her chair and pushed her way past the crowd, head down. Edwards took over at the terminal.

▪ FEC IN FOCUS.

ADOLF HERE—WAITING FOR THROGS.

APPENDIX

This Appendix could have been called "Ardean Science and Technology" except that a great deal of speculation and theorizing by Earth scientists has been added to it. The speculations which began with our first simulation class not only resulted finally in the 2DWORLD program and our contact with Arde, but indirectly created a great deal of communication with scientists and laypersons all over the world, each with something interesting to contribute to our overall picture of the facts and possibilities for science and technology in a two-dimensional universe. In this way our original speculations and theories have been multiplied a hundredfold and, although much of the new scientific material is too technical to include here, much is not. Perhaps some future, more specialized publication will include what it has been necessary to leave out of the present book.

The Appendix consists of six sections, each of which takes the reader somewhat further afield in physics, chemistry, planetary science, biology, astronomy, and technology. I have been guided in my selection of observations and speculations not only by considerations of space and technical difficulty, but by the desire to make clearer to inquiring minds the reasons for certain structures and phenomena which we encountered on Arde. For example, some readers may have realized that much was left out in our discussion of Ardean biology: how are food and hrabx transported through seemingly opaque tissues? These and many other details are explained in the section on biology.

PHYSICS

There could be no better illustration of the need to supplement facts with speculations than the case of Planiversal atoms: although we learned a certain amount of Punizlan quantum theory at the institute, Yendred was not equipped to discuss atomic structure with the scientists there. Neither, for that matter, were we. But it has been possible in the meantime to construct a picture of the Planiversal hydrogen atom based on Punizlan quantum theory.

In our universe, there are four quantum numbers which specify the state of an atom, but in the Planiverse there are only three:

n: the principle quantum number
l: the orbital angular momentum number
m_s: the spin quantum number

Like our own quantum theory, Punizlan theory allows atoms and other Planiversal particles to have only certain energies, all discretely separated from each other and called "quanta." For example, the principal quantum number summarizes the total energy of an atom and this energy can take on only the values of $n = 0,1,2,3. \ldots$, and so on. Likewise, the orbital angular momentum number, which describes the atom's energy of rotation, also has only $l = 0,1,2,3. \ldots$ as its possible values. The spin quantum number, however, is much simpler than these. It may have only two values, $m_s = +1$ or $m_s = -1$, depending on whether the direction of spin is clockwise or counterclockwise.

Equipped with such basic information and solving the two-dimensional Schroedinger equation, Paul Reiser, a physicist from Gloucester, Massachusetts, has obtained the following computer map of electron densities for a Planiversal hydrogen atom.

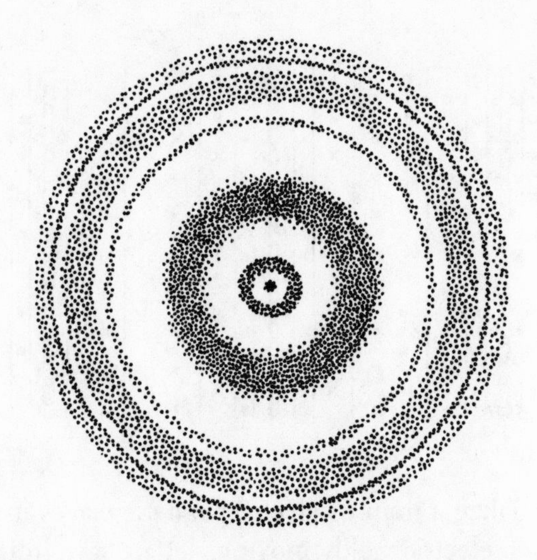

The shaded areas represent the probability of finding the hydrogen atom's single electron in the neighborhood of the nucleus; denser areas represent higher probabilities, and more diffuse areas represent lower ones. In blank areas, the single electron is never found. This particular atom has total energy $n = 3$ and orbital angular momentum $l = 1$. At the very center of the atom is a tiny nucleus.

Every atom and particle in the Planiverse spins either clockwise or counterclockwise. In our universe it is possible to convert one such atom into another merely by rotating it through 180 degrees in our space. But this cannot be done with Planiversal atoms. The two form completely separate populations, and we can only guess how this affects the various interactions of Planiversal matter.

Another fascinating difference between our own universe and the Planiverse is the complete lack of magnetic force in the latter space. Although magnets exist on Arde, they do not attract or repel each other. They interact only with electrons. A number of Earth scientists, including physicists Richard Lapidus of the Stevens Institute in Hoboken, New Jersey, and Ya'akov Stein of Jerusalem, have explored the implications of such an electromagnetic environment and have even obtained versions of Maxwell's equations applicable there. For example, since we know that electromagnetic waves exist in the Planiverse, the following representation of them is perfectly reasonable and almost certainly in accord with the descriptions of Punizlan scientists.

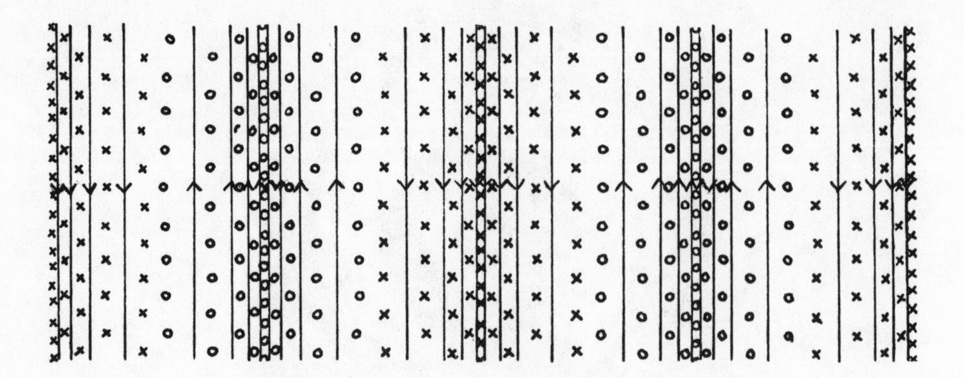

We may visualize a train of electromagnetic waves in the Planiverse as a succession of electric fields moving, let us say, from left to right. Superimposed on each electric field is a magnetic field. As it passes a given point, each electric field builds up to a peak strength and then dies away, to be replaced by an electric field pointing in the opposite direction. Up-pointing electric fields are accompanied by O-type magnetic fields, and down-pointing electric fields are paired with X-type magnetic fields. Naturally, the magnetic fields oscillate along with their electric companions.

Through a two-dimensional vacuum such an electromagnetic wave train will propagate itself endlessly at the Planiversal velocity of light. Farther from the source, these waves grow weaker, like Planiversal gravity.

We knew, or at any rate suspected, even before our contacts with the Planiverse that gravity in a two-dimensional universe would behave differently from our own. As explained in the box entitled "Diminishing Energy" (page 133), gravity in the Planiverse decreases at distance d to a strength proportional to $1/d$, whereas gravity in our own universe diminished by the rule $1/d^2$. In other words, Planiversal gravity weakens with distance much more gradually than our own.

One strange consequence of this law of Planiversal gravity is that a spaceship can never "escape" from an isolated Planiversal planet. On Earth it is only necessary for a rocketcraft to develop a certain speed called the "escape velocity" in order to evade forever the clutches of our gravity. Once this speed has been exceeded, our ship may turn off its motor and continue coasting outward into space as long as it wishes. On Arde, there is no escape velocity. No matter how fast an Ardean rocket ship leaves

Arde and no matter how far away it gets, as soon as the motor is turned off the ship slows down, stops, and begins to fall back toward Arde.

The strength of Planiversal gravity may explain why the Punizlans have never explored very far into the Planiverse beyond Arde. So far they have limited themselves to orbital flights. On the other hand, if they could get close enough to the planet Nagas, they would reach a point where the gravity of Nagas would dominate that of Arde.

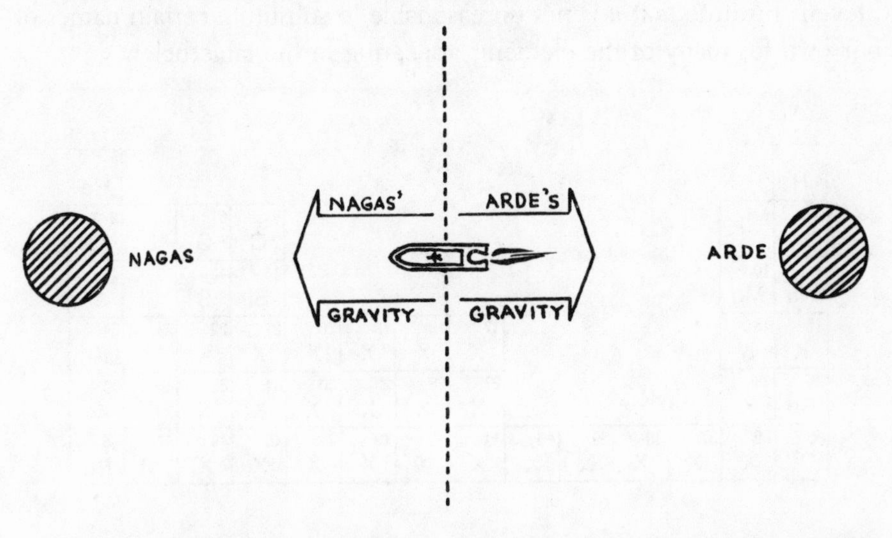

Although I committed the horrendous blunder of setting the scientist Tba Shrin on the trail of relativity theory, I am led to wonder just what sort of relativity theory she will eventually discover. Strangely enough, our own relativity theory predicts no gravity in the Planiverse at all! This was kindly pointed out to me by the astrophysicist Richard Gott of Princeton University. It all has to do with the fact that in Einsteinian relativity the "energy momentum tensor" and the "Riemannian curvature tensor" both have the same number of components in the Planiverse; there is no room for gravity, so to speak. Yet gravity there is! The only hint of a resolution to this paradox comes in a paper by Karel Kuchar of the University of Utah, who points out that there *would* be gravity within a two-dimensional matter field, as though the Planiverse were filled with some kind of impalpable fluid or ether. On this score the Punizlan scientists were silent, but then fish know nothing of water.

CHEMISTRY

In this field we know less and have speculated less than in any other. Although we possess a Planiversal table of the elements, we have very little solid information about Planiversal compounds and chemistry generally.

Punizlan names for their chemical elements are outlandish and unpronounceable. However, I am assured by the chemist Ernest Robb of the Stevens Institute that it is not unreasonable to substitute certain names of our own for many of the elements appearing in the table below.

1 H													2 He
3 Li	4 B									5 C	6 O	7 F	8 Ne
9 Na	10 Mg									11 Si	12 S	13 Cl	14 Ar
15 K	16 X					17 X	18 X	19 X	20 X	21 X	22 X	23 Br	24 Kr
25 Rb	26 X					27 X	28 X	29 X	30 X	31 X	32 X	33 I	34 Xe
35 Cs	36 X	37 X	38 X	39 X	40 X	41 X	42 X	43 X	44 X	45 X	46 X	47 At	48 Rn

For example, the element C (carbon) occupies a point just short of halfway across the second row of the Punizlan table, a position occupied by our own carbon atom. In a geometric sense, moreover, Planiversal carbon with its three valence electrons corresponds nicely to our own carbon with four valence electrons. In the Planiverse, CH_3 would form a triangle, while in the universe CH_4 forms a tetrahedron; these are analogous figures in their respective spaces.

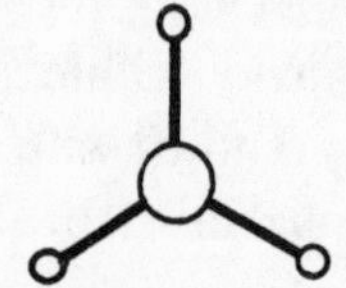

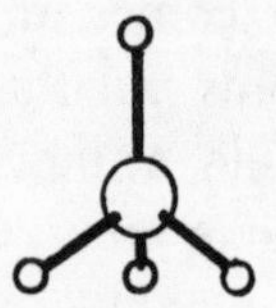

As in our own table of the elements, horizontal position indicates the number of electrons in each shell of the Planiversal atom. Planiversal hydrogen (H) has one electron and helium (He) has two. Two electrons are sufficient to fill the innermost electron shell both here and in the Planiverse, but thereafter the resemblance ends. The next row of our table has eight elements because the next shell in our universe can have up to eight electrons. But the next shell of Planiversal atoms may only hold up six electrons.

The first atom in the second row of the Planiversal table is called "lithium" (Li) by us since, like our lithium, it has just one electron in its second shell. Similarly, Planiversal "fluorine" (F) lacks only one electron, as does ours. Neon (Ne) has a complete second shell and reacts with nothing, either in its Planiversal or universal version. We have thus felt safe in labeling the elements down either side of the Planiversal table with the names of corresponding elements from our universe. All the other elements are labeled X in recognition of their uncertain chemistry. Of these elements we can only say that towards the bottom of the table they almost certainly tend to become unstable and radioactive.

What do Planiversal compounds look like? Does "H_2O" in the Planiverse turn out to be the stuff that fills Fiddib Har?

Virtually all we have is the table but, surely, much can be made of it. For with these few notes the Planiversal symphony is played out, from the high, brilliant arpeggios of biochemistry to the basso continuo of the galaxies themselves!

PLANETARY SCIENCE

This subject includes what we normally call earth sciences (geology, meteorology, and so on) widened to the study of planets in general—including Arde. Having already described the weather systems and surface geography of the planet, it is time to plunge into the planet's interior.

Punizlan scientists have not been lacking in evidence that beneath Ajem Kollosh and beneath the bed of Fiddib Har lies a vast reservoir of molten rock which extends, as far as they can tell, to the very center of the planet.

More than a thousand years ago, a volcano erupted in Vanizla.

About twenty kilometers from Fiddib Har a great fissure opened and red-hot lava bubbled out of it. Simultaneously, according to the ancient records, there came a great rumble and all the land between the sea and the crack began to sink. There was no cinder or lava cone such as volcanoes on Earth construct, merely a great cliff from which the awestruck Vanizlans could look down (one at a time) on a vast pool of lava which extended all the way to the sea, where billows of steam rose in the distance.

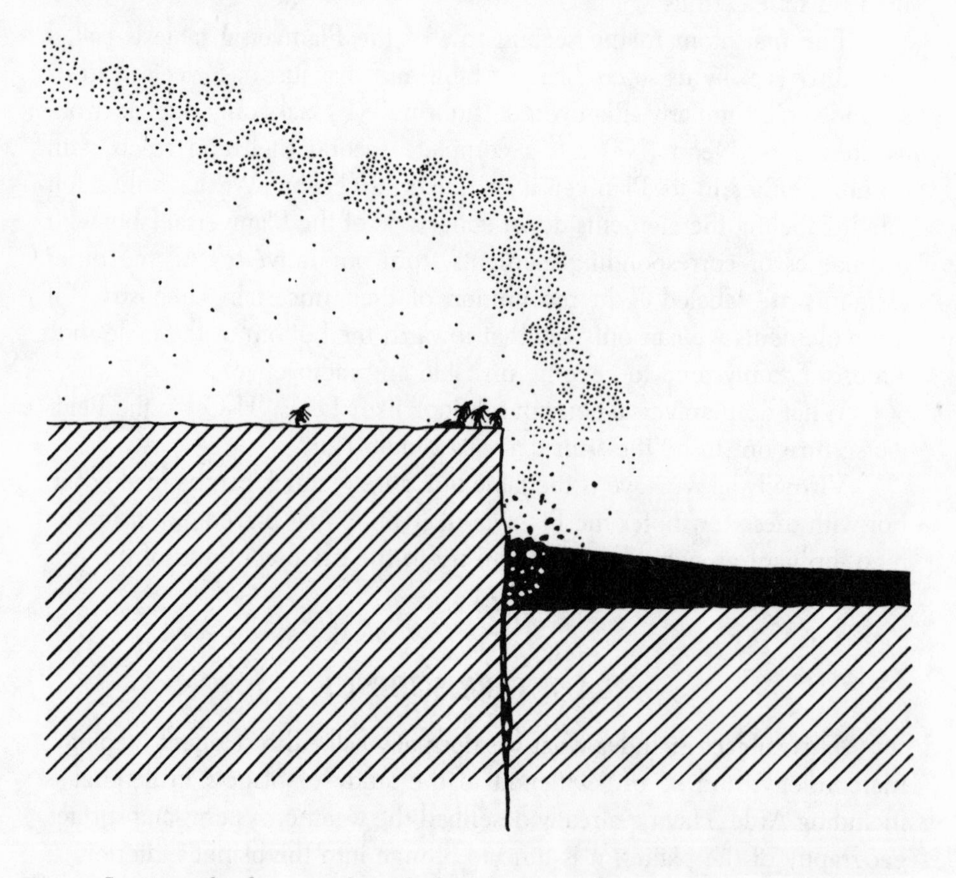

In time the lava pool cooled and new sediments were swept by the rivers of Arde into the newly formed lowland. It probably took no more than a hundred years for the filling-in process to create once again a smooth and continuous surface all the way to Fiddib Har.

In the years of their exploration of the ocean, the Punizlans had encountered a volcano near the middle of Fiddib Har. Not only did this volcano build a "cone" up to the surface of the ocean, but the Punizlan oceanographers discovered a whole chain of extinct undersea volcanoes from one shore of Arde's single ocean to the other.

These and other observations were pieced together at the Punizlan Institute and a theory of strip tectonics, so to speak, soon emerged. In this theory, both Ajem Kollosh and the floor of Fiddib Har together form the crust of the planet and the crust literally floats on the magma which slowly circulates within the planet's interior.

New crust forms at a point near the middle of the ocean where hot, upwelling magma comes very close to the surface. Lighter minerals rise slowly to the very top of the ocean floor, where they cool and replace the crust which is gradually spreading east and west under the inexorable drag of the magma beneath. Occasionally, hot spots develop adjacent to this zone of sea-floor spreading and a volcano erupts its way to the surface of the ocean. New volcanoes appear only every 200,000 or 300,000 years, as far as Punizlan scientists can determine. Nevertheless, in the space of geological time, this accounts perfectly for the chain of seamounts which extends to either shore.

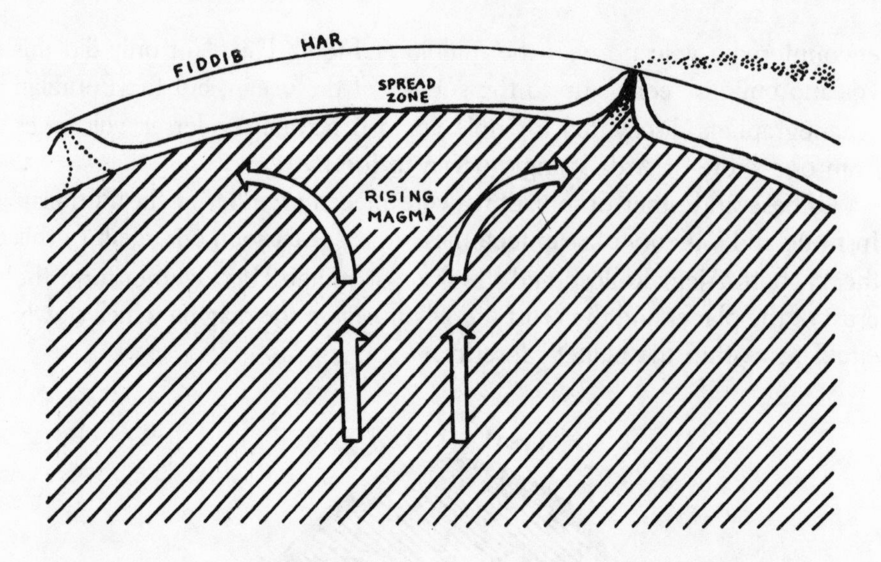

Over a period of several million years the oceanic crust is slowly rafted from the zone of sea-floor spreading to the continent of Ajem Kollosh. The volcanoes ride along on this crust until they encounter the continental mass and are dragged into a trench, where the oceanic crust is pulled down into the magma once again. There is friction as this crust slides below the end of Ajem Kollosh and occasional Ardequakes result from it. The rock of the crust is dissolved in the magma, rejoins the general circulation, and might well reappear in a few million years at the zone of sea-floor spreading.

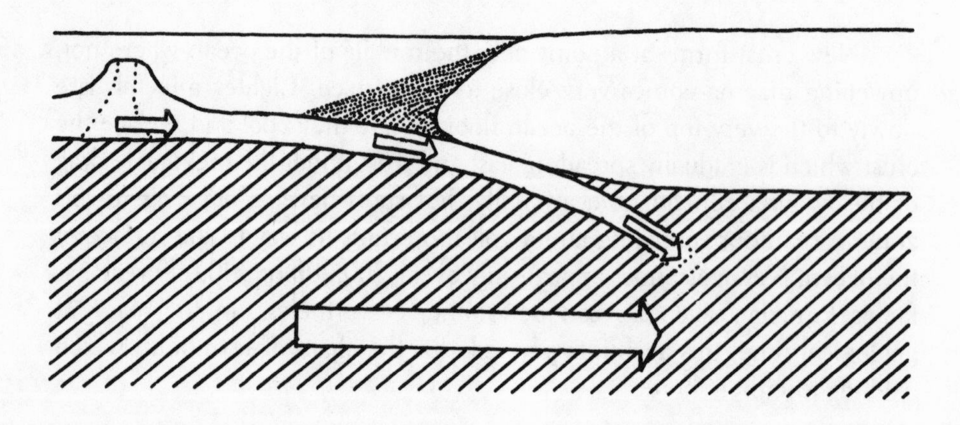

Besides this crust/magma cycle, there is another rock cycle which accounts for the ceaselessly sliding sediments of Ajem Kollosh: the pressure of the two oceanic crusts sliding beneath the two ends of the continent creates a lateral pressure which causes it to bulge in the middle. The bulge is nothing other than Dahl Radam, and great pressure cracks near its upper surface fracture the rock into boulders. Other forces such as cold and heat crack the boulders into rocks, and thus begins the endless supply of gravel, sand, and mud which works its way toward the sea.

When soil reaches the sea, it forms a huge rampart which extends down into the trench. There, it is slowly dragged into the jaws of the subduction zone and remelted into magma. What the continent loses in rocky material from Dahl Radam it occasionally regains in one of two ways: (a) an undersea volcano gets wedged into the subduction zone and becomes glued to the continental mass; (b) lava from a land volcano is added to the surface of Ajem Kollosh. In the latter case, the reason for the sinking of land seaward of the volcano is easily seen.

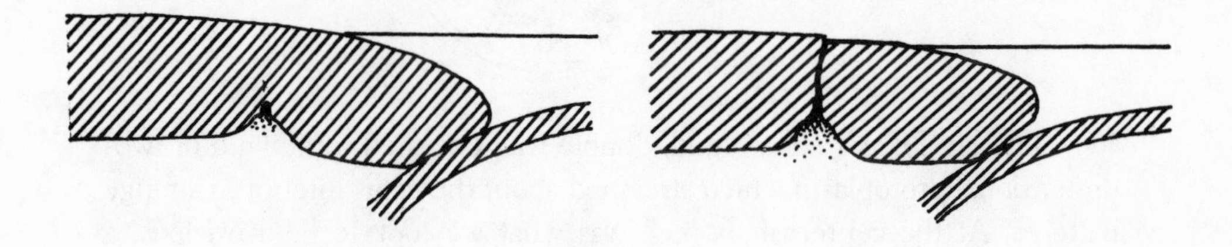

Those familiar with the rudiments of plate tectonics as applied to Earth by our own geologists will have already seen how similar the Punizlan theory is to our own. But Earth has several continents, while Arde has only one. Here again the incredible simplicity of Arde reemerges as a scientific theme. But Arde's lone continent results not from its two-dimensionality but from something a little more subtle. As the geologist Tuzo Wilson of the University of Toronto pointed out many years ago, Earth probably has several continents because it has a solid core which subdivides an otherwise simple circulation pattern into many smaller ones. Arde, however, has no solid core. As far as either Punizlan scientists (or our own) can determine, no two-dimensional matter can be solid at the temperatures which prevail at the center of Arde.

BIOLOGY

The life forms of Arde are composed of cells which lie at the lowest limit of the 2DWORLD program's resolution. However, we have discovered and recorded free-living single-celled life forms which are quite a bit larger than the cells composing Ardean plants and animals. We found a tiny creature (which we later learned was called the Hi Tikek) living within the small, liquid space between four soil particles. This perfect, miniature thing swam languidly about its little world digesting microscopic spots.

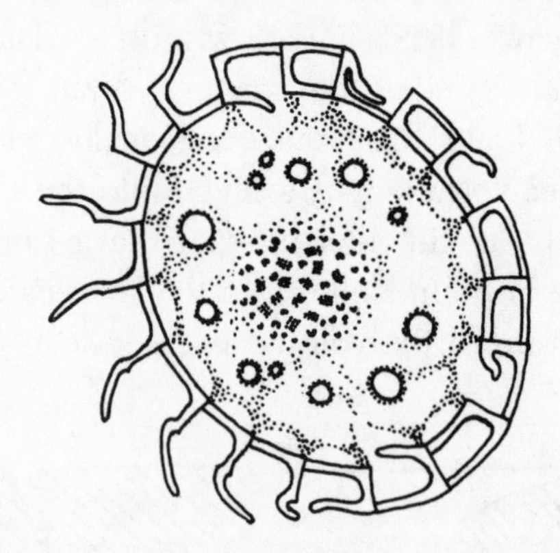

The Hi Tikek's chambered, double wall enclosed some kind of two-dimensional protoplasm which streamed about the cell's interior in strange patterns. At the center of the cell was what we took to be a nucleus, a diffuse collection of fibrous bodies spread out, perhaps, to allow biochemistry in the plane.

The cell wall, a circlet of U-shaped chambers, held the cell together while permitting the transfer of materials into and out of the cell. Each U-shaped chamber consisted of two arms, one of which always held on to the wall of an adjacent unit. How the attachment was made we could only guess. A kind of biochemical glue? Microscopic molecular hooks? Should both arms of any chamber have simultaneously released their hold, the Hi Tikek would have broken up before our eyes, releasing its precious protoplasm into the water and dying.

The outer arms of the chambers were used both for swimming and for catching food. They appeared capable of coordinated motion, and the manner in which they brought small particles into their respective chambers was fascinating: an arm would fold inward into its chamber and lengthen itself against the opposite arm until, suddenly, it would flash outward when triggered by the presence of a food particle. The inrush of water into the chamber frequently carried the particle with it and, when the outer arm had reattached itself, the inner arm would detach and allow the particle to drift into the cell's interior.

We know now that if we had watched the Hi Tikek in its tiny domain long enough, we would have seen it swim about with increasing lethargy as the dissolved hrabx and food particles were used up by its life processes. Gradually its watery space would have grown exhausted and stale with expelled wastes. The Hi Tikek would then become dormant, awaiting the next river to release it from its tomb.

As we scanned the damp soil in search of similar creatures, we came upon other Hi Tikeks but nothing else. Is this the only kind of microscopic creature which inhabits Arde? We doubt that. Nevertheless, such low diversity would be very characteristic of the planet.

One of the Hi Tikeks we thus encountered did not move, and all of its chambers were closed. It appeared to be in the act of dividing down the middle.

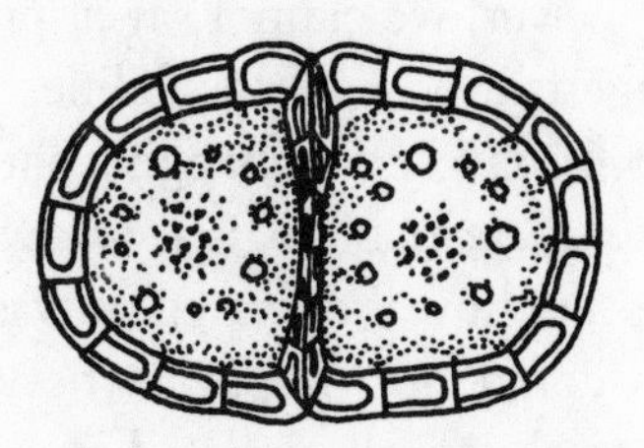

Some kind of filament had evidently divided the cell into two halves and new chambers were being elaborated along its length, from the outside inward. Unfortunately, we saw no other stages of cell division (if that's what it was) during the brief hour we spent scanning the soil.

From the biologists at the Punizlan Institute, we had obtained a description of how material flows from cell to cell in Ardean tissue. This description only made sense if we assumed that such cells were enclosed in the same circuit of chambers as the Hi Tikek. Our hypothetical reconstruction of this mechanism is shown in the figure below where two cells meet at a common boundary, their chambers precisely aligned.

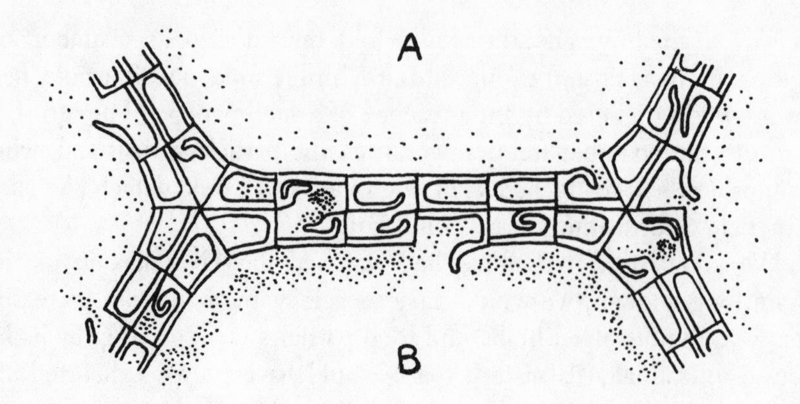

By the nearly simultaneous rolling back of the outer chamber arms, the fluid contents of two adjacent chambers are quickly exchanged. This mutual withdrawal of arms creates a miniature current which swirls within the newly created space. When the fluid has circulated a half-turn, the outer arms abruptly close and the inner arms wave the chamber contents into the interiors of their respective cells. In this way a concentration of nutrients in cell A would quickly diffuse into cell B. To what degree the diffusion is selective, however, we cannot say at present. The basic mechanism nevertheless suggests a host of possibilities, and is almost certainly the only way Ardean cells have of exchanging their contents.

In questioning the Ardean biologists, we were especially eager to learn about nerve tissue and how it operated. It was already clear to us that conducting nerve cells could not propagate nerve impulses along their walls because these were electrically isolated from each other, unlike the membranes of our own nerve cells.

If we have got it right, nerve cells are rather long and arranged into a kind of triangular tresswork with a very thin membrane separating adjacent cells. I am assured by the biologist David Clark, recently a graduate student at the University of Oregon, that this is a perfectly workable arrangement of nerve cells and that *action potentials* (nerve impulses) are most probably generated along the common membrane, zigging and zagging their way to or from the brain. Again, we are forced to use a hypothetical reconstruction.

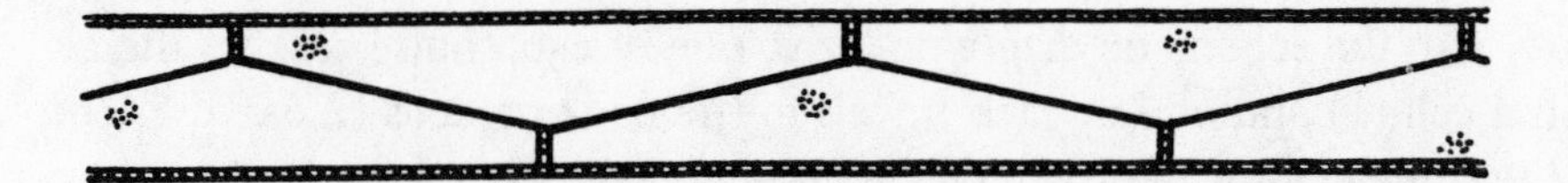

Do Ardean nerve cells work like ours? Does the thin membrane pump ions from one cell to its neighbor in order to generate a voltage potential? Perhaps we shall never know the answers to these questions.

Another crucial feature which we learned from Punizlan biologists was how nerve impulses cross each other within the space of an Ardean brain. A little triad of cells manages this feat quite nicely but would be quite unnecessary if by some magic we could get two nerve fibers to cross each other in a two-dimensional space.

Two nerve pulses, traveling along fibers *a* and *b,* first encounter a split in the fibers, and two copies of each pulse are sent along the paired fibers.

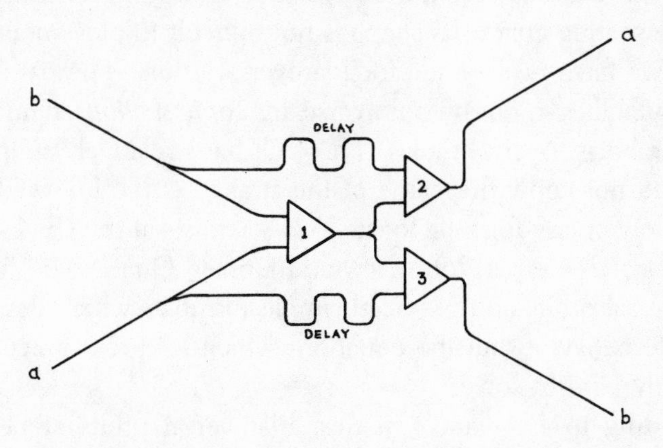

In the schematic diagram above, one of each pulse reaches the central cell (1) and one of each travels to the flanking cells (2 or 3). Each of the three cells in the triad will generate a new pulse along its output fiber if and only if exactly one pulse arrives along either input fiber; if two pulses arrive simultaneously, it will not fire. Following this very simple rule, it is not hard to see that the same pattern of pulses arriving on input fiber *a* will leave again a split second later along the output fiber with the same label. The same thing holds for fiber *b*.

So it is that whatever Ardean neurons are capable of, whether they are like our own or not, they do not suffer for lack of communication with their colleagues. The absence of a third dimension places no essential restriction on their interconnections; it was once argued by the scientist G. J. Whitrow more than twenty years ago that intelligence would be impossible in a two-dimensional space for this very reason. Obviously, the Planiverse has much to teach us!

ASTRONOMY

There seem to be only two planets pursuing their weaving orbits around Shems, namely Arde and Nagas. The scarcity of planets in the Shems system may be typical of the Planiverse generally. Large planetary systems might be unstable because of the far-reaching effects of Planiversal gravity.

As we have already seen, the orbit of Arde around Shems is a beautiful self-intersecting curve. Its shape is not difficult to plot on a computer, but as yet we have no formula for Planiversal orbits. The paths followed by heavenly bodies in our own universe are *conic sections,* namely circles, ellipses, parabolas, or hyperbolas. These all have relatively simple formulas. But I do not know the name of the strange curve followed by Arde, nor do I know of any formula for it. Two scientists at the IBM Thomas J. Watson Research Center, John Lew and Donald Quarles, Jr., have classified Planiversal orbits and provided simple formulas which describe their approximate behavior, but the equations which lead to a precise formula are devilishly hard to solve.

According to Lew and Quarles, Planiversal orbits are completely determined by three parameters, the phase angle, size, and eccentricity.

The first two parameters are not essential to this discussion—only the eccentricity, which alone determines the shape of orbits. One way of describing eccentricity is to specify the ratio of seasonal year to sidereal year.

The *closed* orbits are those which retrace their own path after a finite number of revolutions around Shems.

Except for the circle, each of the orbits shown above has a definite number of lobes, two, three, four, and so on. The *nonclosed orbits* never retrace themselves and are rather difficult to draw since their drawing can never be completed! Arde traces just such an orbit around Shems.

There are larger two-dimensional phenomena to classify than mere orbits. Earlier I stated that the Planiverse itself was like a sphere, but that was hardly more than an assumption on my part and something less than a firm conclusion on the part of Punizlan astronomers. In fact, the possibilities for the ultimate topology of the Planiverse are infinite. For example, if the Planiverse is what mathematicians call *orientable,* then it is either a sphere or it is like a doughnut with one or more holes in it.

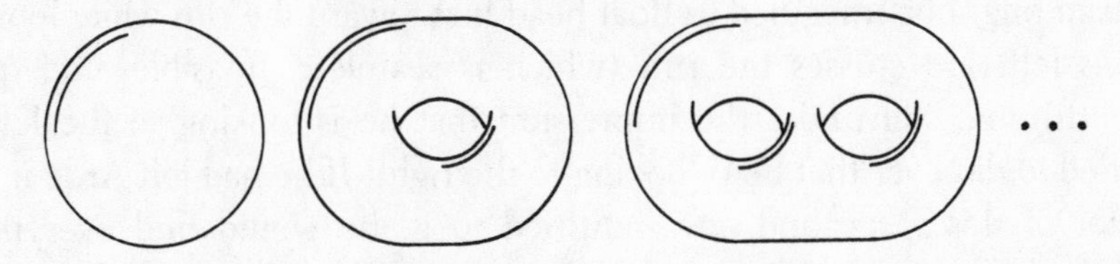

If the Planiverse is *non-orientable,* on the other hand, it would be a weird place indeed. The simplest non-orientable two-dimensional universe would be a *projective plane,* a space curved back upon itself in such a peculiar way that an Ardean in such a universe could go on a long space voyage and come back with his left and right sides reversed. The next diagram shows a disk-shaped mini-universe which is about to become a projective plane.

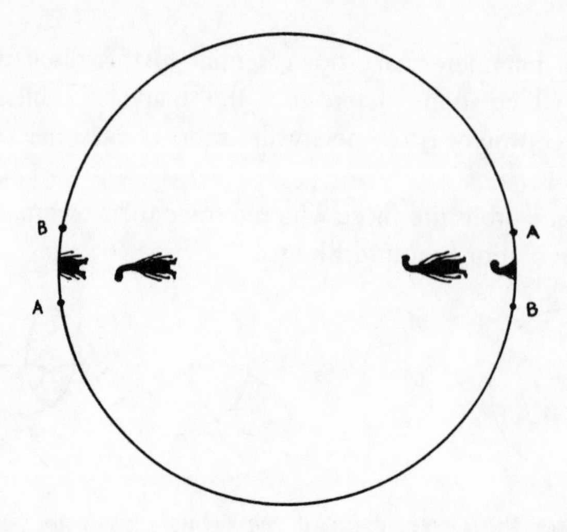

Ignoring the Ardeans for a moment, imagine that the two points labeled *A* are attached together and then the points labeled *B.* In fact, attach each point on the rim of the disk to the opposite side of the rim and do this all the way around the rim, as though sewing the disk up so that when finished, it no longer has a rim or any sort of boundary. The resulting object was once called "Fortunato's purse" by Lewis Carroll.* Such a space is very difficult for (three-dimensional) humans to imagine for the simple reason that it cannot be constructed in three dimensions. Instead, four dimensions are required. Nevertheless, it is not difficult for humans to appreciate what goes on in such a place. For example, an Ardean might be imagined to float head first toward the rim while looking to his left. He crosses the rim (which is seamless, invisible, and quite smooth) and, still under the impression that he is looking to the left, is startled to discover that he is looking to the right! If he had left Arde in the center of this space and now returned to it, he would find everything reversed. For example, he could no longer read Punizlan books but would find Vanizlan books perfectly legible.

But such bizarre possibilities for the Planiverse are hardly more realistic to imagine than our own universe being non-orientable.

Besides topology, there is the geometry of the Planiverse to consider.

* Because everything outside this curious surface is also inside it. Considered as a "purse," it contains all the world's wealth.

Questions of geometry hinge largely upon whether the Planiverse is closed or not. The usual examples of open two-dimensional universes are the infinite plane and an infinite saddle-shaped space. But if we are to take the Punizlan astronomers' statements at face value, such universes must be excluded. They do not "on themselves curve back."

Astronomers and cosmologists on Earth have been trying for decades to discover whether our own universe is closed or not and, indeed, whether it will ever collapse back upon itself. Fortunately, the second question is very easy to answer in the case of the Planiverse. Like our own universe, the Planiverse seems to have started with a big bang. It will surely all collapse in the end back to the nothing whence it came. Planiversal gravity will enforce this ultimate collapse, for even as the Planiversal galaxies flee each other, continuing the momentum of the big bang, they will slow down, stop, and fall back together just as surely as the Ardean rocket ship could never escape Arde into free space. It will all come to an end someday, all the galaxies, Shems and Arde, Punizlans and Vanizlans, crushed into a mathematical point, the zero-dimensional beginning and end of existence itself.

TECHNOLOGY

Engineers analyze all mechanisms constructed on Earth as composed of certain basic and indecomposable simple machines like the inclined plane, the lever, and so on. One may do something similar for the mechanisms of Arde by listing all the elementary components from which the larger and more complex machines have been constructed.

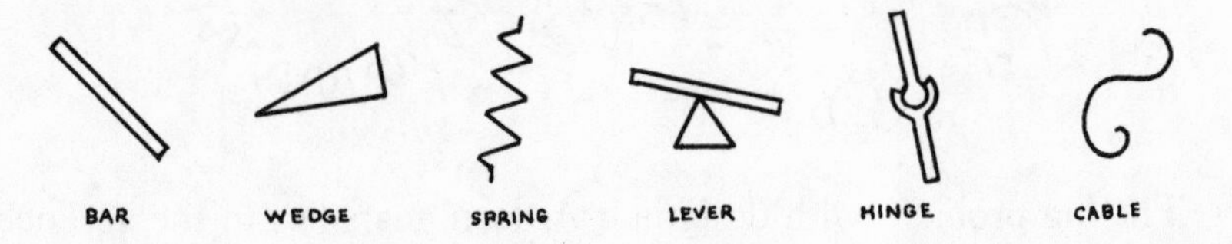

Of course, these simple machines come in many shapes and sizes. For example, a spring does not have to be shaped like a zigzag: it could be a leaf spring like the one operating the cam in the steam engine. The hinge, moreover, comes in a variety of forms, some of them quite sophisticated. For example, the hinge shown below opens just as widely or even wider than the standard hinge and is much stronger.

Strangely enough, the same hinge is used here on Earth! This was pointed out to me by Norman Allen, a technician in Ottawa. The advanced Ardean hinge is basically the cross section of a hinge used on aluminum toolchests.

Another coincidence which recently caught my attention was the strong resemblance between the gearing system built by one of the Rdidn brothers and a set of cogs designed by the British mathematical physicist Roger Penrose of the Mathematical Institute at Oxford.

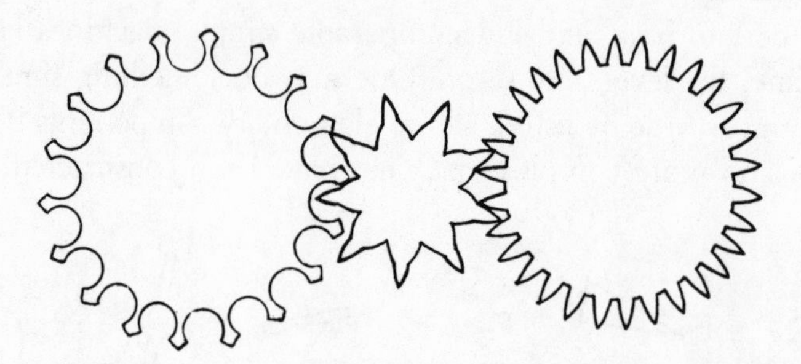

The big problem with designing Ardean gears lies in the absence of axles. Typically, here on Earth, a high rate of turn is converted to a low one, or vice versa, by the presence of two gears on a single axle. This cannot be done on Arde because axles are impossible or, in any event, inaccessible. The purpose of Ardean gears, in fact, is not to vary the rate of revolution in going from one gear to the next, but to vary the surface speed of the gears. This means, quite literally, the speed with which the gear's teeth (or some point on the teeth) travel around the center of the gear. No matter how many gears of the ordinary sort we might wish to assemble in the Planiverse, however, the surface speed of the first gear in a train is always the same as the last one—assuming of course that each gear can somehow be fixed to rotate in a single plane. It therefore requires quite extraordinary gears to accomplish the required change in the relative speeds of their surfaces (or teeth). This is precisely what the gears above do.

The large gear on the left and the gear in the center both have two radii, in effect. The whole trick by which the gearing down of surface speed is accomplished lies in the way the large gear's outer radius teeth engage the central gear's inner radius teeth. The central gear's outer teeth, on the other hand, barely touch the large gear on the left because of the special hollows scooped out between the large gear's teeth. However, the same outer teeth of the central gear engage the teeth of the gear on the right. So it is that the surface speed of the outer radius of the left gear becomes the surface speed of the central gear's inner radius. This speed is increased, in effect, by the fact that the central gear's outer radius is traveling faster than its inner one. After all, it is the central gear's outer radius which engages the large gear on the right.

Nearly every house and building which Yendred entered during his journey across Punizla had a strange-looking device affixed to one of its walls. Inside a narrow, vertical box a series of what can only be described as "quarter-gears" rocked back and forth. The gears at the bottom of the box rocked quickly while those near the top moved very slowly, the motion of the topmost gear being nearly imperceptible. It did not take us long to realize that these were clocks. A string from the topmost gear ran outside the box to a weight hanging down one side of the clock. The Ardeans told the time by observing the position of this weight as it slowly moved up the side of the box, revealing one number after another.

A spring connecting the arm of the topmost gear to the framework of the clock provides its motive power. To "wind" the clock, an Ardean must wait until noon (the Punizlan 0-hour) and then remove the framework holding the topmost gear in order to disengage the gear from the mechanism and pull it toward him as far as it will come. This action loads the spring. Replacing the topmost gear framework, the Ardean has now set the clock but is unable to witness its simple and beautiful operation.

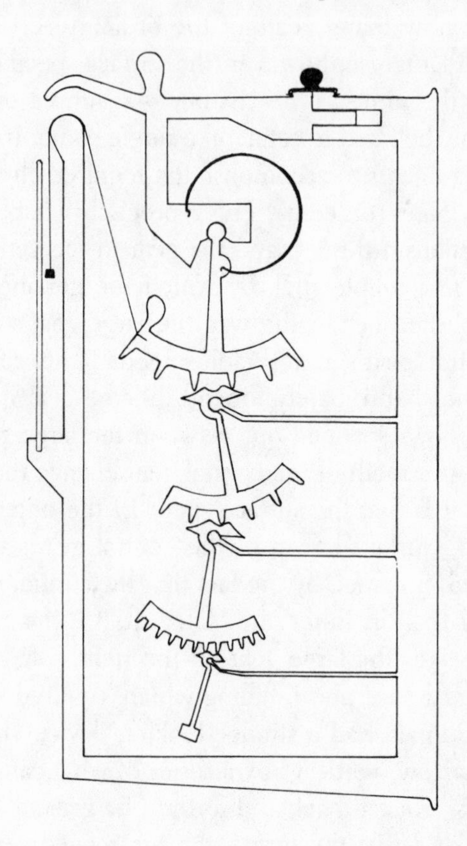

Each quarter-gear engages a rocking cam which is attached to the next quarter-gear below it. As the teeth of the gear pass over the rocking cam, it wags the attached quarter-gear back and forth, however slowly. The force of the spring is transmitted in this way to the lowest gear, which rocks a pendulum back and forth. So it is that for each twenty-four swings of the pendulum, the lowest gear swings back and forth once. For each six swings of the lowest gear, the next higher gear swings once. In the clock shown above, the topmost gear will complete one left-to-right motion when the pendulum has completed $24 \times 6 \times 6 = 864$ swings. Actual Punizlan clocks have three more intermediate gears than are shown above, so that by the time one Arde-day has gone by the pendulum has completed something less than 186,624 swings and the clock will have to be rewound.

Although the Planiverse itself will remain forever inaccessible to us, there is a sense in which we three-dimensional creatures may participate in its two-dimensional reality: each of the mechanical devices shown in these pages, even the clock above and the steam engine described earlier, can actually be built in our world! Using the drawings in this book as templates, the various moving parts of the machines may be cut out of any thin, reasonably strong material and assembled into the actual device. In case I have got some of the more critical proportions wrong, it would be advisable first to make a working model in cardboard in order to adjust and trim the various components. Then one could move to some more durable material such as plastic or steel.

In the chapter on the Punizlan Institute, I described the NAND gate used by Punizlan computers without saying anything about how such gates could be assembled into more complex circuits. Any textbook on computer circuit design will show how this can be done and there is no point in duplicating such material here. However, it may not be immediately obvious how to get signals to cross each other in a Planiversal computer. In order to construct a crossover, the Punizlans first assemble an XOR gate from four NAND gates, each symbolized by a semicircle with a small circle attached to it.

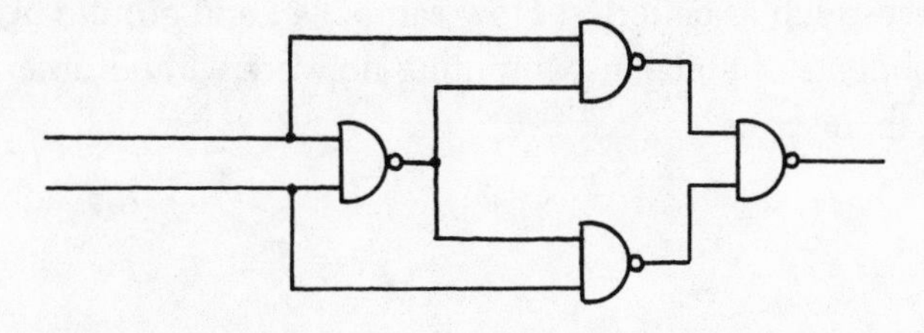

The XOR gate computes the eXclusive-OR of its two inputs, x and y: the output of this gate, symbolized by $x \oplus y$, is high if and only if either x or y (but not both) is high. Otherwise its output is low.

From three XOR gates, each symbolized by a shield-shaped symbol, a crossover circuit is then easy to construct.

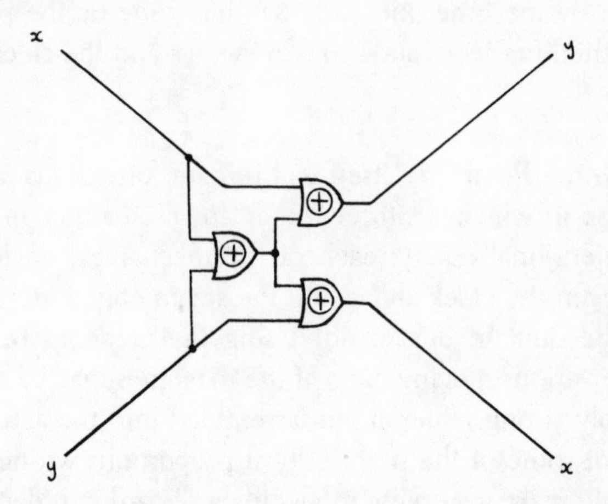

This configuration of XOR gates (in reality a slightly complicated little circuit of 12 NAND gates) has two input lines and two output lines. The output line labeled *x* will carry exactly the same signal (high or low) as the input line labeled *x*, independently of whatever signal is carried by the other input line labeled *y*. The same statement can be made about the lines labeled *y*. So it is that a signal arriving on the upper left line can leave on the lower right one without being affected by the other signal.

Carlo Sequin of the University of California at Berkeley has carried out a theoretical study of other ways in which two-dimensional computing might be managed, including the use of two-dimensional relays and bifilar networks. He has found it very challenging to avoid the use of localized power sources such as batteries. How can power and ground be connected to all components of a planar computing network without interfering with the signals themselves?

ACKNOWLEDGMENTS

If the Planiverse did not already exist, it would surely have been invented, at least in the sense that many people, myself included, had already been speculating about two-dimensional worlds. First among these was Edwin A. Abbott, an English clergyman who wrote the now famous *Flatland* in 1884. Some years later, in 1907, Charles Hinton, an American logician, wrote *An Episode of Flatland* which reorganized Abbott's tabletop world into the somewhat more logical disk planet, Astria. Much later, in 1965, Dionys Burger, a Dutch physicist, published *Sphereland,* which attempted to reconcile both Abbott's and Hinton's worlds and then to use the resulting two-dimensional universe to illustrate the curvature of space.

It was in May 1977 that I was reading a popular work on cosmology and came across the familiar analogy which describes the expansion of our own three-dimensional universe in terms of a balloon whose two-dimensional skin continually expands. This led me to speculate about whether it would be possible, taking the balloon analogy literally, for a two-dimensional universe actually to exist. What sort of physics and chemistry would it have? What sort of life forms?

In spare time which was becoming increasingly scarcer, I began to compile speculations and theories about two-dimensional physics. As a member of faculty in a very busy and growing computer science department, I already had less leisure time than my colleagues in other departments but I made the most of what I had and, in 1979, published a small monograph called *Two-Dimensional Science and Technology,* following this, early in 1980, with a revised edition of that work. In the summer of 1980, Martin Gardner published an article about these writings in the July issue of *Scientific American* and the 2,000-odd requests from readers exhausted my supply of books. They have never been reprinted.

Among the few people I had corresponded with while writing *Two-Dimensional Science and Technology* was Jef Raskin of Apple Computers, Inc., who suggested the basic design used in the rocket plane appearing both in the monograph and in this book. I am also indebted to my colleague Andrew Szilard of the University of Western Ontario, who in 1979 plotted some of the preliminary orbits for a hypothetical two-dimensional planet circling its sun under an inverse-linear force law.

Because of the enormous response to the *Scientific American* article, I felt encouraged to organize a further publication called *A Symposium on Two-Dimensional Science and Technology.* Published during my sabbatical at Oxford in 1981, this book featured articles and notes by scientists and laypersons alike on everything from physics to two-dimensional pianos. I wish to acknowledge several of the contributions to *A Symposium on Two-Dimensional Science and Technology* because some of the ideas, phenomena, and devices described in foregoing pages are based upon these articles. In doing so, some of the names which have just appeared in the Appendix will be repeated. In particular, it was Norman Allen who sent me an aluminum hinge which happens to work quite nicely in two dimensions. David Clark wrote two articles on two-dimensional biology which contained designs for a cell wall, DNA, and nerve cells, among other things. Richard Lapidus wrote several articles about electricity and magnetism, hydrogen atoms, black body radiation, heat capacity, and the theory of gases, and collaborated with his colleague Ernest Robb in producing the two-dimensional periodic table in the Appendix. John Lew and David Quarles wrote a definitive paper on two-dimensional orbits, not only plotting several but specifying for which parameters orbits are possible. Professor Penrose designed the two-dimensional gears at the Institute in Oxford between bouts with black holes and other space-time singularities. Paul Reiser not only analyzed planar hydrogen atoms but investigated two-dimensional electrons and electrodynamics, as well as designing ingenious Planiversal electric motors which are simply too complicated for the Ardeans ever to invent! In his article on two-dimensional computing, Carlo Sequin described a number of approaches to the problem of simultaneously routing power, ground, and logical signals in planar logical devices. Finally, Ya'akov Stein wrote an article on Maxwell's equations in two dimensions.

Among those not mentioned in the Appendix but also deserving credit are Sergio Aragón of the Universidad del Valle Guatemala, George Marx of Roland Eötvös University in Budapest, and Timothy Robinson of Christchurch, New Zealand, all of whom constructed two-dimensional tables of the elements for the *Symposium.* It was Bobby Clayton of Bothell, Washington, who designed the boat in which Yendred and his father went fishing. I have elaborated somewhat on Clayton's original design. Alex Comfort, the well-known psychiatrist and adviser on human relations, contributed a paper on plant and animal designs. I have used his notion of low-lying plants. Edmund Hellfrich of Allentown, Pennsylvania, and Kenneth Knowlton of Bell Labs in Murray Hill, New Jersey, both designed water delivery systems. Unfortunately, these could not be made to work properly in an Ardean context and the Punizlans have been (so far) doomed to the most primitive facilities. Glen Lesins, a graduate student at the University of Toronto, speculated on Ardean weather patterns, confirming and precisely recasting earlier speculations of my own. I used Lesins' analysis in my description of the high and low which daily circle Arde. John Mamin of Berkeley, California, contributed a fascinating paper on the Dirac delta function, noting that it developed a curious tail when propagated as a two-dimensional wave. Robert Munafo of Warren, Rhode Island, gets much of the credit for the Planiversal clock which appears in the Appendix. I redesigned his escapement mechanism in what I hope is an improvement to the clock. Saul Roe of Los Angeles devised a steam-operated elevator which I modified slightly for use in the Punizlan battery factory. Richard Wells, a maker of musical instruments at the University of Waterloo, designed a two-dimensional piano, a variation of which appears at the concert attended by Yendred and his girlfriend, Na.

Some of the minor devices appearing in my story about Arde were suggested by numerous people in what is probably one of the more remarkable instances of simultaneous invention in recent history. These include the unworkable wheels of the Rdidn brothers' cars, Ardean books, hinges, springs, and various other simple mechanisms. The closest thing to a functional Arde car, namely the one with the wraparound wheels, was first drawn to my attention by Wallace Riley, a technical writer from San Francisco. Since I received his letter, ten other people have independently discovered it.

Before continuing the history of how the present book came into being, it would only be fair to draw attention to some of my own colleagues and students at the University of Western Ontario who contributed in various ways to it. Paddy Nerenberg and Stan Deakin of our applied mathematics department became interested in the phenomena of the "Dirac tails" described by Munafo, and Deakin, with the help of two graduate students, developed a computer solution for general wave propagation in two dimensions. The various distortions undergone by different sources is still far from clear, but it was Deakin's students Peter Kvas and Stephen Sawchuck who drew my attention to the brief frequency change which accompanies the onset of sounds in Arde. Much earlier than this, even before the publication of the first monograph in 1974, my colleagues Andrew Szilard and Julian Davies had developed a certain interest in the Planiverse. Szilard plotted some preliminary orbits for two-dimensional planets, and Davies made one of the early forays into electromagnetic theory in the plane. Harold Johnson, a former graduate student in computer science, suggested the difficulty which Ardeans would have in lifting a plank.

Finally, there is a chain of events, each represented by a single person, which led more or less directly to the publication of this book. My special thanks go to Martin Gardner for his brilliant and fascinating article in *Scientific American* in July 1980. It was this article which lifted our speculation about two-dimensional science and technology onto a new plateau by bringing it to the attention of a much wider public and thereby enhancing the quality of contributions to the subject. The next event in the chain occurred in the fall of 1981 when a press release, written by Susan Boyd of *The Western News,* our campus newspaper, received international attention and resulted in magazine articles and television stories publicizing our two-dimensional world, which at that time carried the name Astria. In particular, Sharon Begley of *Newsweek* magazine wrote a story for the "Ideas" section which caught the attention of many people in the publishing world. Besides my thanks to Boyd and Begley, I must next thank my agent, Nancy Colbert, who helped to develop the idea of a book and to sell it to a publisher. The publisher selected was Pocket Books, whose editor Marnie Hagmann of Washington Square Press worked untiringly to cut and polish this rough stone. She and Ann Patty of Poseidon Press also deserve my thanks. Along the way, typists Elizabeth Masson and Tanya Spruyt have provided excellent service and useful advice. Finally, Pat Dewdney, my wife, read the later drafts and made some crucial suggestions which made all the difference between a merely good book and one which we all hope is excellent.

ABOUT THE AUTHOR

ALEXANDER KEEWATIN DEWDNEY teaches Computer Science at the University of Western Ontario in London, Canada, the city of his birth. A mathematician by training, he studied as an undergraduate at the University of Western Ontario, and did his graduate work at the University of Michigan and the University of Waterloo, where he received a Ph.D. in 1974. Dewdney's major interests for the last 20 years have been Discrete Mathematics and Theoretical Computer Science, fields in which he has published extensively. However, it was reading Edwin Abbott's *Flatland*, written 1884, that led him to speculate seriously on possible two-dimensional universes. Dewdney lives in London, Canada, with his wife Pat, a library scientist, and their son Jonathan.